Springer Series on
ATOMIC, OPTICAL, AND PLASMA PHYSICS   48

Springer Series on
ATOMIC, OPTICAL, AND PLASMA PHYSICS

The Springer Series on Atomic, Optical, and Plasma Physics covers in a comprehensive manner theory and experiment in the entire field of atoms and molecules and their interaction with electromagnetic radiation. Books in the series provide a rich source of new ideas and techniques with wide applications in fields such as chemistry, materials science, astrophysics, surface science, plasma technology, advanced optics, aeronomy, and engineering. Laser physics is a particular connecting theme that has provided much of the continuing impetus for new developments in the field. The purpose of the series is to cover the gap between standard undergraduate textbooks and the research literature with emphasis on the fundamental ideas, methods, techniques, and results in the field.

Vols. 20–46 of the former Springer Series on Atoms and Plasmas are listed at the end of the book

F. Grossmann

Theoretical Femtosecond Physics

Atoms and Molecules in Strong Laser Fields

With 91 Figures

 Springer

sep/ac
phys.

Dr. Frank Grossmann
Technische Universität Dresden, Institut für Theoretische Physik
01062 Dresden, Germany
E-mail: frank@physik.tu-dresden.de

Springer Series on Atomic, Optical, and Plasma Physics   ISSN 1615-5653

ISBN 978-3-540-77896-7   e-ISBN 978-3-540-77897-4

Library of Congress Control Number: 2008927183

Typesetting and prodcution: SPI Publisher Services
Cover concept: eStudio Calmar Steinen
Cover design: WMX Design GmbH, Heidelberg

SPIN 12112470   57/3180/SPI
Printed on acid-free paper

9 8 7 6 5 4 3 2 1

springer.com

To the memory of my father
Hans Grossmann

Preface

The development of modern pulsed lasers with power densities larger than $10^{16} \mathrm{W\,cm}^{-2}$ and with very short pulse duration in the femtosecond regime enables experimentalists to study elementary processes such as chemical reactions and excitation mechanisms in different areas in physics in the time domain. In parallel to the experimental investigations, analytical and numerical studies of laser-driven atoms and molecules with a limited number of degrees of freedom are performed. These theoretical investigations have led to the prediction and/or the explanation of a large variety of partly counter-intuitive phenomena. Among those are the generation of high harmonics using laser-excited atoms or molecules, the ionization of atoms above the continuum threshold, the stabilization of atoms against ionization in very strong fields, counter-intuitive pulse sequences to selectively populate vibrational states in molecules and, last but not least, the control of chemical reactions by specially tailored laser pulses.

This book originated from a course, that I have been giving on a regular basis since 2000 for advanced undergraduate and graduate students at Technische Universität Dresden. It offers a theoretically oriented approach to the field of laser-driven atomic and molecular systems and requires some knowledge of basic classical and quantum mechanics courses as well as of classical electrodynamics. The book has two introductory chapters in Part I that pave the way for the applications in Part II. Part I and also Chap. 3 of Part II contain of textbook knowledge that is needed to understand the rest of the book. The material presented in the last two chapters is close to the recent literature. I have chosen only such works, however, that deal with fundamental concepts and are based on simple model calculations. A biased and incomplete list of references is given at the end of the chapters, preceded by some notes and hints for further reading. For those readers who are interested in some computational details, these are given in the appendices at the end of the corresponding chapters. Furthermore, at several places throughout the text, exercises are placed, whose independent solution allows a deeper understanding of the material presented.

In Chap. 1, we start with a short introduction into the foundations of the laser. We concentrate especially on those aspects of pulsed lasers that will be important for the theoretical investigations in Part II of the book.

The next fundamental chapter is devoted to the non-relativistic time-dependent Schrödinger equation. In the case of lasers of up to atomic field strengths, this equation allows the theoretical description of the phenomena we want to investigate in Part II. Analytical as well as numerical methods to solve the time-dependent Schrödinger equation are thus in the focus of Chap. 2. Throughout the book, to keep the approach as simple as possible, we touch the topic of correlated many particle dynamics only where necessary and concentrate on the description of electronic as well as nuclear dynamics, with the help of models with as few degrees of freedom as possible. The contents and the presentation of Chap. 2 are inspired by the excellent new textbook by David Tannor, *Introduction to Quantum Mechanics: A Time-Dependent Perspective*, which hopefully will start a "revolution" in the way quantum mechanics is taught in the future.

The second part of the book, starting with Chap. 3, contains a collection of equivalent ways to couple a charged particle to a classical electromagnetic field. As the basic postulate, we use the principle of minimal coupling. By using unitary transformations, one can then either derive the length form or the Kramers–Henneberger form of the coupling. As first examples of laser–matter interaction, we study the dynamics of (structure-less) two-level systems in laser fields. Phenomena like Rabi oscillations of the occupation probability, occurring there, will be encountered off and on in the remainder of the book. Furthermore, also the fundamental so-called rotating wave approximation will be discussed for the first time in this context.

Selected examples of laser–matter interaction in atomic physics are reviewed in Chap. 4. Here, we concentrate on the phenomena of ionization and high harmonic generation of a single electron in a Coulomb potential of possibly reduced dimensionality. It turns out that a perturbation theoretic approach would not be suited to understand most of the phenomena presented in this chapter. Thus, the numerical wavepacket methods that were in the focus of Chap. 2 will find their first application.

The next step in the direction of higher complexity of the dynamics will be taken in Chap. 5. Here, we deal with laser-driven systems in molecular and chemical physics. The simplest molecule, the hydrogen molecular ion, H_2^+, will serve as a vehicle to understand some of the basic concepts of molecular physics, such as electronic potential surfaces. In the following, the full numerical solution of the coupled electron nuclear problem of H_2^+ in a monochromatic laser field will be reviewed. After discussing the fundamental Born–Oppenheimer approximation, in the rest of the chapter, we then assume that the solution of the electronic many-body problem is at our disposal in the form of analytically or numerically given potential energy surfaces. After a short digression on nuclear motion on a single electronic surface, and the discussion of a simple two coupled surfaces problem, we then review some modern applications in the fields of femtosecond spectroscopy, optimal control theory, and quantum information processing under the foregoing assumption.

At this point, I thank the students at TU Dresden who have attended my lectures. They have inspired me enormously, through their intense collaboration, during the lectures, as well as during the exercise classes. This has motivated me to consider the material presented here again and again and the students have thus contributed substantially to the improvement of the manuscript. Also the hospitality of the Max-Planck-Institute for the Physics of Complex Systems, that offered me the opportunity to attend and run several conferences in the field was very important to shape my understanding presented here. Furthermore, I express my deep gratitude to Jan-Michael Rost and Rüdiger Schmidt for their continuous availability for discussions and for long-term collaboration. Moreover, I am grateful to Peter Hänggi for the introduction to the field of driven-quantum systems during my PhD work with him, and to Eric Heller for opening the world of time-dependent semiclassics to me. In addition, I have benefitted from innumerable discussions with and valuable advice of former members of the Theoretical Quantum Dynamics Group in Freiburg, especially Gernot Alber, Richard Dehnen, Volker Engel, Christoph Meier, Gerd van de Sand, and Gerhard Stock. Furthermore, former and present members of the Theoretical Atomic and Molecular Physics Group at the Institute of Theoretical Physics of TU Dresden and the Finite Systems Department at the MPIPKS in Dresden have helped shape my understanding. Among many others these are Andreas Becker, Agapi Emmanouilidou, Celal Harabati, Anatole Kenfack, Thomas Kunert, Ulf Saalmann, and Mathias Uhlmann. For helping me by answering specific questions or supplying information and valuable graphs, I thank Wolfgang Schleich, Jan Werschnik, Matthias Wollenhaupt and Shuhei Yoshida. For advice and help with respect to graphics issues, I thank Arnd Bäcker and Werner Koch. Finally, I am indebted to David Tannor, who supplied me with preliminary versions of his book at a very early stage and thus helped shape the presentation here to a substantial degree. The focus of David's book on a time-dependent view of quantum phenomena is an absolute necessity if one wants to study laser-driven systems.

Dresden, May 2008   *Frank Grossmann*

Contents

Part I

Prerequisites

A Short Introduction to Laser Physics

To study the influence of light on the dynamics of an atom or a molecule experimentally, laser light sources are used most frequently. This is due to the fact that lasers have well-defined properties. The theory of the laser dates back to the 1950s and 1960s of the twentieth century and by now, 50 years later is textbook material. In this introductory chapter, we start by recapitulating some basic notions of laser theory, which will be needed to understand later chapters.

More recently, experimentalists have been focusing on pulsed mode operation of lasers with pulse lengths of the order of femtoseconds, allowing for time-resolved measurements. At the end of this chapter, we therefore put together some aspects of pulsed lasers that are important for their application to atomic and molecular systems.

1.1 The Einstein Coefficients

Laser activity may occur in the case of nonequilibrium, as we will see later. Before dealing with this situation, let us start by considering the case of equilibrium between the radiation field and an ensemble of atoms in the walls of a cavity. This will lead to the Einstein derivation of Planck's radiation law.

The atoms will be described in the framework of Bohr's model of the atom, allowing the electron to occupy only discrete energy levels. For the derivation of the radiation law, the consideration of just two of those levels is sufficient. They shall be indexed by 1 and 2 and shall be populated such that for the total number of atoms

$$N = N_1 + N_2 \tag{1.1}$$

holds. This means that N_2 of the atoms are in the excited state with energy E_2 and N_1 atoms are in the ground state with energy E_1. Transitions between the states shall be possible by emission or absorption of photons of the appropriate energy. The following processes can be distinguished:

- Absorption of light leading to a transition rate

$$\frac{\mathrm{d}N_2}{\mathrm{d}t}\bigg|_{\text{abs}} = \rho N_1 B_{12} \tag{1.2}$$

from the ground to the excited state.
- Induced (or stimulated) emission of light leading to a transition rate

$$\frac{\mathrm{d}N_1}{\mathrm{d}t}\bigg|_{\text{emin}} = \rho N_2 B_{21} \tag{1.3}$$

for the population change of the ground state.
- Spontaneous emission of light leading to a rate

$$\frac{\mathrm{d}N_1}{\mathrm{d}t}\bigg|_{\text{emsp}} = N_2 A \tag{1.4}$$

which amounts to a further increase of the ground state population.

The first two processes are proportional to the energy density ρ of the radiation field with the constants B_{12}, respectively, B_{21}. The process of spontaneous emission does not depend on the external field and is proportional to A. These coefficients are called Einstein's A- and B-coefficients.

In thermal equilibrium, the rate of transition from level 1 to 2 has to equal that from 2 to 1, leading to the stationarity condition

$$N_1 B_{12} \rho = N_2 B_{21} \rho + N_2 A. \tag{1.5}$$

This equation can be resolved for the energy density ρ leading to

$$\rho = (N_1 B_{12}/(N_2 B_{21}) - 1)^{-1} A/B_{21}. \tag{1.6}$$

Furthermore, in thermal equilibrium, the ratio of populations is given by the Boltzmann factor according to

$$N_1/N_2 = \exp\left\{-\frac{E_1 - E_2}{kT}\right\} \tag{1.7}$$

with the temperature T and the Boltzmann constant k.

As $T \to \infty$ also $\rho \to \infty$, and we can conclude that the B-coefficients have to be identical $B_{12} = B_{21} = B$. Using Bohr's postulate

$$E_2 - E_1 = h\nu, \tag{1.8}$$

where ν is the frequency of the light, we can conclude from (1.6) that

$$\rho = \left(\exp\left\{\frac{h\nu}{kT}\right\} - 1\right)^{-1} A/B. \tag{1.9}$$

holds. The ratio of Einstein coefficients A/B can now be determined by comparing the formula above with the Rayleigh–Jeans law

$$\rho(\nu) = \left(8\pi/c^3\right)\nu^2 kT, \tag{1.10}$$

which is a very good approximation in the case of low frequencies (see Fig. 1.1). One then arrives at

$$A/B = \left(8\pi/c^3\right)h\nu^3 =: D(\nu)h\nu \tag{1.11}$$

for the ratio, where $D(\nu)\mathrm{d}\nu = 8\pi\nu^2/c^3\mathrm{d}\nu$ is the number of possible waves in the frequency interval from ν to $\nu+\mathrm{d}\nu$ in a cavity of unit volume [1]. Inserting this result into (1.9) yields Planck's radiation law

$$\rho\mathrm{d}\nu = (D(\nu)\mathrm{d}\nu)h\nu \left(\exp\left\{\frac{h\nu}{kT}\right\} - 1\right)^{-1}. \tag{1.12}$$

The last factor in this expression is the number of photons with which a certain wave is occupied. As a function of the wavelength, Fig. 1.1 shows a comparison of Planck's law with the two laws only valid in the limits of either long or short wavelength. These are the Rayleigh–Jeans and Wien's law, respectively.

In the case of nonequilibrium, an extension of the formalism just reviewed leads to the fundamentals of laser theory, as we will see in the following. The explicit calculation of the Einstein B-coefficient shall be postponed until Chap. 3.

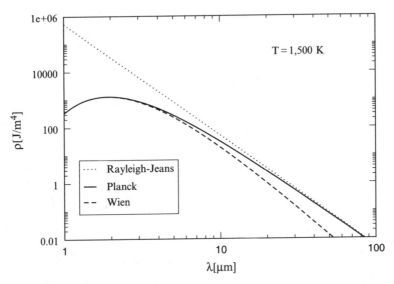

Fig. 1.1. Energy density (per wavelength interval) as a function of wavelength for different radiation laws at a temperature of $T = 1{,}500\,\mathrm{K}$

1.2 Fundamentals of the Laser

The derivation of laser activity can be done in a crude way by again considering the populations of two levels between which the laser transition occurs. The atoms are driven out of equilibrium by pumping and are interacting with a fixed frequency light field with photon number n (considered to be a continuous variable in the following) in a resonator [2].

First, we consider the processes leading to a change in the populations. In addition to the ones introduced in Sect. 1.1, these are pump (or gain) and loss processes. We concentrate on laser activity and therefore spontaneous emission can be neglected for the time being. Secondly, the realization of the laser process, requiring more than a bare two-level system will be shortly discussed.

1.2.1 Elementary Laser Theory

In close analogy to the Einstein coefficients for the induced transition rates, coefficients can be defined that fulfill $W_{ij} = W_{ji} = W$ leading to an induced emission rate of $(N_2 - N_1)Wn$.[1] Including the gain and loss processes, depicted in Fig. 1.2, the rate equations

$$\frac{dN_1(t)}{dt} = \gamma_{12}N_2 - \Gamma N_1 + (N_2 - N_1)Wn, \qquad (1.13)$$

$$\frac{dN_2(t)}{dt} = \Gamma N_1 - \gamma_{12}N_2 - (N_2 - N_1)Wn, \qquad (1.14)$$

emerge. Subtracting the first from the second equation leads to a rate equation for the difference $D = N_2 - N_1$, which is also referred to as population inversion

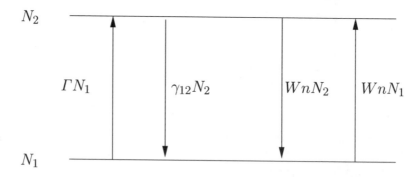

Fig. 1.2. Two-level system with elementary transitions (from the *left* to the *right*): Pump process, loss processes (e.g., by radiationless transitions), induced emission and absorption; spontaneous emission is not considered; adapted from [2]

---

[1] Note that in the previous section the rate was proportional to ρ and here it is proportional to the dimensionless variable n; we therefore have to use a different symbol for the coefficients.

$$\frac{\mathrm{d}D}{\mathrm{d}t} = -2WnD - \frac{1}{T_1}(D - D_0). \tag{1.15}$$

Here the definitions of the unsaturated inversion $D_0 = N(\Gamma - \gamma_{12})/(\Gamma + \gamma_{12})$, which will become clear below, and the relaxation time $T_1 = (\Gamma + \gamma_{12})^{-1}$ have been introduced. Including loss effects of the optical cavity via a parameter t_{cav}, the rate equation for the photon number

$$\frac{\mathrm{d}n}{\mathrm{d}t} = WnD - \frac{n}{t_{\mathrm{cav}}}, \tag{1.16}$$

follows, where the first term is due to the increase of radiation by stimulated processes and the effect of spontaneous emission has been neglected. Equation (1.15) for the inversion together with (1.16) for the photon number are a simplified version of the full quantum mechanical laser equations, allowing one to understand some basic laser properties [2,3].

For an amplification of the light field to occur by starting from a low initial photon number n_0 with unsaturated inversion D_0, the right-hand side of (1.16) has to be larger than zero. For reasons of simplicity, let us here just consider the steady state defined by

$$\frac{\mathrm{d}n}{\mathrm{d}t} = 0 \qquad \frac{\mathrm{d}D}{\mathrm{d}t} = 0, \tag{1.17}$$

however. For the inversion we get

$$D = D_0/(1 + 2T_1Wn), \tag{1.18}$$

i.e., a reduction for a finite photon number as compared to the unsaturated value D_0. The photon number in the steady state follows from

$$n\left(\frac{WD_0}{1 + 2T_1Wn} - \frac{1}{t_{\mathrm{cav}}}\right) = 0, \tag{1.19}$$

leading to two different solutions:

(1) $n_0 = 0$
(2) $n_0 = (D_0 - D_{\mathrm{thr}})\frac{t_{\mathrm{cav}}}{2T_1}$

In order for the nontrivial solution to be larger than zero, the inversion has to be larger than a threshold value $D_{\mathrm{thr}} = 1/(Wt_{\mathrm{cav}})$. As a function of D_0, the transition from the trivial solution to the one with a finite number of photons is depicted in Fig. 1.3.

In principle, laser theory has to be formulated quantum theoretically. This is done e.g., in [2]. There the transition from a standard light source to a laser above threshold is explained in a consistent framework. For large photon numbers one finds the phenomenon of anti-bunching, i.e., the photons leave the cavity equidistantly. The corresponding laser light has a constant amplitude. Therefore in the applications part of this book, we will assume that the field can be described classically by using a sinusoidal oscillation.

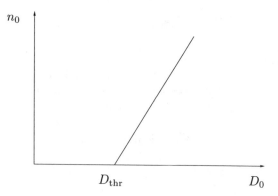

Fig. 1.3. Steady state photon number versus unsaturated inversion

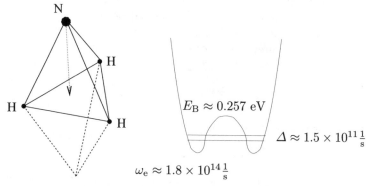

Fig. 1.4. Umbrella mode of NH_3 indicated by the *arrow* and schematic double well (frequency of oscillation around the minima ω_e, tunneling frequency Δ, and barrier height E_B are indicated) with the two levels (their separation is vastly exaggerated for reasons of better visibility) used for the maser process

1.2.2 Realization of the Laser Principle

As we have just seen, nonequilibrium, characterized by population inversion, is crucial for operating a laser. Since the invention of the first maser[2] it has been shown that inversion can be achieved in many different ways. A small collection of possibilities (including also the microwave case) will now be discussed.

The Ammonia Maser

In the NH_3-maser [4], the umbrella mode (see Fig. 1.4) leads to a double well potential and thus quantum mechanically tunneling is possible. A corresponding doublet of levels in the double well exists, which is used for the

---

[2] Maser stands for "Microwave amplification by stimulated emission of radiation".

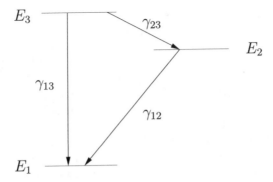

Fig. 1.5. Three-level system of the ruby laser with the metastable level E_2

maser process. Inversion is created by separating the molecules in the upper level from the ones in the lower level by using the quadratic Stark effect in an inhomogeneous electric field.

This principle cannot be applied in the optical (i.e. laser) case, however, since typically $h\nu \gg kT$ at optical frequencies and therefore $N_2 \ll N_1$. Increasing the number of atoms in the upper level via pumping is therefore necessary.

The Ruby Laser

To achieve inversion in a laser, more than two levels are needed. Solid-state lasers like the three-level ruby laser [5] are pumped optically. Lasing is then done out of the metastable level E_2, shown in Fig. 1.5. By considering just the pumping and the loss terms in the rate equations for the three-level system one can show that

$$\Gamma > \gamma_{12}\left(1 + \frac{\gamma_{13}}{\gamma_{23}}\right) \tag{1.20}$$

has to hold for $N_2 > N_1$, which can be fulfilled with moderate pumping under the conditions $\gamma_{12} \ll \gamma_{13}$ and $\gamma_{23} \gg \gamma_{13}$ [3].

Exercise 1.1 *Consider an extension of the rate equations to the three-level case and neglect the induced terms. Under which condition for the pumping rate Γ can population inversion between the second and first level be achieved?*

Other Types of Lasers

Other types of lasers are gas lasers, in which the laser active medium is pumped by collisions with electrons or atoms and the transitions can be either electronic (He–Ne laser) or ro-vibronic ones (CO_2 laser).

In addition, there are semiconductor-based lasers, dye lasers, excimer lasers, to name but a few. Their working principles are described in some detail in [2, 6]. Another special laser type is the free electron laser (FEL), where a high speed electron beam is accelerated in a spatially modulated magnetic field and thereby emits coherent light. Recently, the principle of the FEL has been realized in two new large scale experiments. An infrared FEL has e.g., been built in Dresden (Rossendorf) and the FEL FLASH (formerly VUV-FEL) at DESY in Hamburg generates radiation in the soft X-ray regime.

A common principle in the experimental setup of all lasers is the fact that spontaneous emission (being a form of isotropic noise) should be suppressed. This is a difficult task, especially for high frequencies, however, due to the fact that $A \sim B\nu^3$, see (1.11), holds for the Einstein coefficients. Details of the experimental setup as e.g., the quality factor of the cavity have to be considered to understand how temporal fluctuation tend to get washed out, see e.g., [7].

1.3 Pulsed Lasers

Experimentally, lasers have led to a revolution in the way spectroscopy is performed. This is due to the fact that lasers are light sources with well-defined properties. They can be operated continuously in a single mode modus with a fixed or a tunable frequency or in a multi-mode modus [6]. However, more important for the remainder of this book is the possibility to run lasers in a pulsed mode. There the laser only oscillates for a short time span (e.g., some femtoseconds) with the central frequency of the atomic transition that is used.

1.3.1 Frequency Comb

Experimentally, ultrashort laser pulses can be created by using the principle of mode locking, explained in detail e.g., in [6]. We will shortly discuss the superposition of a central mode with side bands, underlying that principle below. The net result is shown in Fig. 1.6, where a train of femtosecond pulses coupled out of a cavity is depicted. Among other possible applications to be

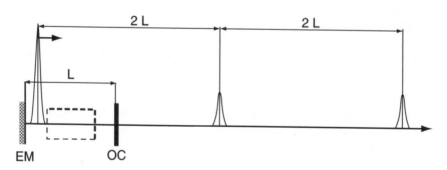

Fig. 1.6. Laser with end mirror (EM), output coupler (OC) and a pulse, propagating between EM and OC and being partially transmitted, from [9]

discussed in detail in later chapters, a pulse train can be used to measure frequencies very precisely [8].

How are the side bands obtained experimentally, and why does their superposition together with the central frequency ν lead to a train of pulses? The first question can be answered by considering the periodic modulation of the inversion with the frequency

$$\delta\nu = c/(2L) = 1/T_{\mathrm{RT}},\tag{1.21}$$

corresponding to the round trip time T_{RT} of the light in the resonator. With the modulator placed at some position inside the cavity, the possible resonator modes with the angular frequencies

$$\omega \pm 2\pi n\delta\nu\tag{1.22}$$

with $\omega = 2\pi\nu$ and $n = 0, 1, 2, 3, \ldots$ are amplified [6]. Peaks at these equidistantly spaced frequencies are called the frequency comb.

To answer the second question, the amplitude of the electric field at a fixed point in space,

$$\mathcal{E}(t) = \sum_{n=-p}^{p} \mathcal{E}_n \cos[(\omega + 2\pi n\delta\nu)t + \varphi_n],\tag{1.23}$$

has to be considered, where $\varphi_n = n\alpha$ are the locked phases. A total of $2p+1$ modes shall have a gain above the threshold value. In the case of $\mathcal{E}_n = \mathcal{E}$ and for $\alpha = 0$ this leads to an intensity of

$$I(t) \sim \mathcal{E}^2 \left| \frac{\sin[(2p+1)\pi\delta\nu t]}{\sin(\pi\delta\nu t)} \right|^2 \cos^2(\omega t).\tag{1.24}$$

Exercise 1.2 *Derive a closed expression for the electric field and the corresponding intensity in the case of the mode locked laser by using the geometric series.*

The intensity of (1.24) contains a term describing a fast oscillation with the central frequency and an envelope function leading to peaks separated by the round trip time $T_{\mathrm{RT}} = 1/\delta\nu$. Furthermore, the pulse length[3] is $T_{\mathrm{p}} \approx 1/\Delta\nu$, with the inverse width parameter $\Delta\nu = (2p+1)\delta\nu$, increasing linearly with the number of participating modes. The intensity as a function of time for three different total numbers of contributing modes is displayed in Fig. 1.7. The peak intensity increases proportional to $(2p+1)^2$, whereas the pulse length decreases with $1/(2p+1)$.

The effect of the pulse generation can also be understood in the photon picture. Those photons passing through the modulator at times where its

---

[3] Defined as the full width at half maximum of the intensity curve.

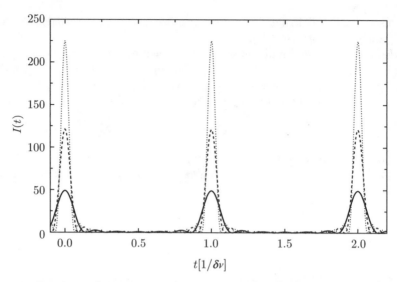

Fig. 1.7. Envelope of the intensity (arbitrary units) of a pulse train as a function of time for the superposition of 7 (*solid line*) 11 (*dashed line*) and 15 modes (*dotted line*)

transmission has a maximum will experience a minimum loss and the corresponding light will be maximally amplified. Enormously high intensities on the order of 10^{16}W cm^{-2} can be generated using the principle of passive mode locking [6]. They prevail only for short times on the order of several femtoseconds, however. Pulses with 6 fs length are nowadays generated with Ti:Sapphire lasers with Kerr lens mode locking and operate at a center wavelength of 800 nm [10]. Only about two oscillations of the field are contained in such a short pulse at those wavelengths. The light is therefore extremely polychromatic. Many further details regarding experimental realization can be found in Chap. 3 of [9].

1.3.2 Carrier Envelope Phase

Let us look at the electric field of the last section in a bit more detail. It consists of an oscillation with the central frequency ω under an envelope and is plotted for a certain choice of parameters in Fig. 1.8.

The parameters in Fig. 1.8 have been chosen such that the peak separation does coincide with a half integer multiple of the period of the fundamental oscillation. This results in the fact that the phase of the fundamental oscillation is different by π, whenever the envelope has reached its next maximum. In general, this phase difference is the so-called carrier envelope phase (CEP) $\Delta\varphi$, and later-on we will frequently adopt the form

$$\mathcal{E}(t) = \mathcal{E}_0 f(t) \cos(\omega(t)t + \Delta\varphi) \tag{1.25}$$

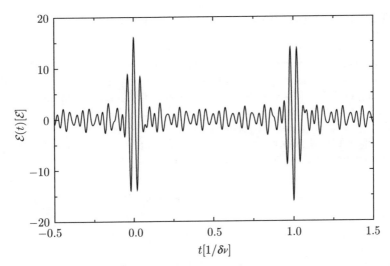

Fig. 1.8. Laser field consisting of the superposition of 17 modes with central frequency $\nu = 4$, $L = 3.0625$, and $\alpha = 0$ (all quantities in arbitrary units) as a function of time

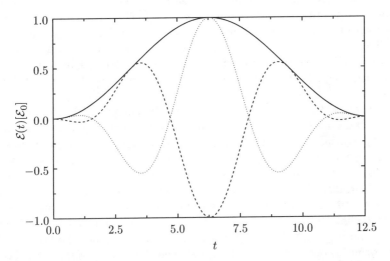

Fig. 1.9. Schematic laser field oscillation under a *single* pulse envelope (in units of $\mathcal{E}_0$) but with two different values of the carrier envelope phase as a function of time in arbitrary units, analogous to the two different pulses depicted in Fig. 1.8

of the laser field with an amplitude $\mathcal{E}_0$ and an envelope function $f(t)$, which is chosen from a large variety of suitable analytic functions. In addition, the frequency ω might be time-dependent and the carrier envelope phase can be varied, leading to tremendous effects as we will see later. In Fig. 1.9, a *single* pulse with oscillations corresponding to two different values of $\Delta\varphi$ is shown.

1.3.3 Husimi Representation of Laser Pulses

As mentioned in Sect. 1.3.2, we will model a laser pulse by an oscillation times a freely chosen pulse envelope. This is reasonable due to the fact that arbitrarily formed laser pulses can be generated experimentally by so-called pulse shapers [9]. We will make use of this fact in Chap. 5 in connection with the control of chemical reactions. In the following, the representation of the frequency content of general laser fields will be discussed.

A very intuitive way to characterize a laser pulse is given by a "windowed" Fourier transformation (or Husimi transformation)

$$F(\tau, \Omega) = \left| \int_{-\infty}^{\infty} dt g(t - \tau) \mathcal{E}(t) e^{-i\Omega(t-\tau)} \right|^2 \tag{1.26}$$

with the window function

$$g(t) = \exp[-t^2/(2\sigma^2)]/\sqrt{2\pi\sigma^2}. \tag{1.27}$$

The function $F(\tau, \Omega)$ depending on a time-like variable, as well as on a frequency is also referred to as a spectrogram. It tells us at which time τ a certain frequency Ω is present in the original signal $\mathcal{E}(t)$. The term frequency resolved optical gating (FROG) is used for a measurement technique of a pulse which is designed by using (1.26) [9, 11]. In the field of molecular spectra, the term vibrogram [12] is used for a quantity which is constructed in a similar way from a time-signal called auto-correlation function, to be defined in the next chapter.

The case of two pulses which are temporally delayed with respect to each other will occur frequently later-on. For such a so-called pump–dump pulse, with slightly different central frequency of the pump versus the dump pulse, a spectrogram is shown in Fig. 1.10. The frequency change and also the temporal delay is clearly visible in the spectrogram. Also the case of a single pulse with a so-called "up chirp" (central frequency increasing as a function of time) or a "down chirp" (central frequency decreasing) are very obvious in a corresponding Husimi plot. To verify this for a simple Gaussian pulse envelope, a Gaussian integral has to be performed. This is by far not the last one that appears in this book and for convenience some Gaussian integrals are collected in Appendix 1.A.

Exercise 1.3 *For the case of a linearly chirped frequency*

$$\omega(t) = \omega_0 \pm \lambda t/2$$

calculate and schematically depict the Husimi transform of the pulsed field

$$\mathcal{E}(t) = \mathcal{E}_0 \exp\left[-\frac{(t - t_0)^2}{2\sigma^2} + i\omega_0(t - t_0) \pm i\frac{\lambda}{2}(t - t_0)^2 \right].$$

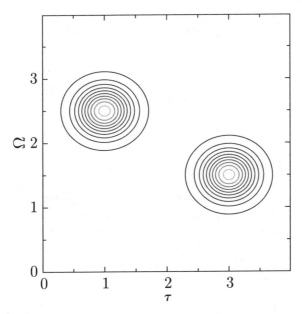

Fig. 1.10. Husimi transform of a pump–dump pulse as a function of τ and Ω in arbitrary units

1.A Some Gaussian Integrals

Throughout this book, Gaussian integrals will be encountered. For complex-valued parameters a and b with $\mathrm{Re}\, a \geq 0$, the following formulae hold:

$$\int_{-\infty}^{\infty} dx \exp\{-ax^2\} = \sqrt{\frac{\pi}{a}}, \tag{1.28}$$

$$\int_{-\infty}^{\infty} dx\, x \exp\{-ax^2\} = 0, \tag{1.29}$$

$$\int_{-\infty}^{\infty} dx\, x^2 \exp\{-ax^2\} = \left(\frac{1}{2a}\right)\sqrt{\frac{\pi}{a}}, \tag{1.30}$$

$$\int_{-\infty}^{\infty} dx \exp\{-ax^2 + bx\} = \sqrt{\frac{\pi}{a}} \exp\left\{\frac{b^2}{4a}\right\}, \tag{1.31}$$

$$\int_{-\infty}^{\infty} dx\, x \exp\{-ax^2 + bx\} = \left(\frac{b}{2a}\right)\sqrt{\frac{\pi}{a}} \exp\left\{\frac{b^2}{4a}\right\}, \tag{1.32}$$

$$\int_{-\infty}^{\infty} dx\, x^2 \exp\{-ax^2 + bx\} = \left(\frac{1}{2a}\right)\left(1 + \frac{b^2}{2a}\right)\sqrt{\frac{\pi}{a}} \exp\left\{\frac{b^2}{4a}\right\}. \tag{1.33}$$

A generalization of one of the formulae given above to the case of a d-dimensional integral that is helpful is

$$\int d^d x \, \exp\{-\boldsymbol{x} \cdot \mathbf{A}\boldsymbol{x} + \boldsymbol{b} \cdot \boldsymbol{x}\} = \sqrt{\frac{\pi^d}{\det \mathbf{A}}} \exp\left\{\frac{1}{4}\boldsymbol{b} \cdot \mathbf{A}^{-1}\boldsymbol{b}\right\}. \qquad (1.34)$$

As in the 1d-case, it can be proven by using a "completion of the square" argument. Furthermore, the convention that non-indication of the boundaries implies integration over the whole range of the independent variables has been used.

Notes and Further Reading

The theory of the laser is treated on the level of the rate equations as well as in its full quantum version in the book by Haken [2] (the first book of the series [1,2] contains the derivation of Plancks's law, that we have followed) and by Shimoda [3]. In these books one can also find a more detailed discussion of the rate equations beyond the steady-state solution, especially concerning the build up of the oscillation.

A lot of information on the experimental aspects of lasers and about mode locking are contained in the book by Demtröder [6]. The handbook article by Wollenhaupt et al. deals with the properties, the creation via mode locking, and the measurement of femtosecond laser pulses [9]. It also contains a long list of additional references. The characterization of short pulses by using FROGs is the topic of [9,11].

References

1. H. Haken, *Licht und Materie Bd. 1: Elemente der Quantenoptik* (BI Wissenschaftsverlag, Mannheim, 1989)
2. H. Haken, *Licht und Materie Bd. 2: Laser* (BI Wissenschaftsverlag, Mannheim, 1994)
3. K. Shimoda, *Introduction to Laser Physics* (Springer, Berlin, 1984)
4. J.P. Gordon, H.J. Zeiger, C.H. Townes, Phys. Rev. **95**, 282 (1954)
5. T.H. Maiman, Phys. Rev. Lett. **4**, 564 (1960)
6. W. Demtröder, *Laser Spectroscopy* (Springer, Berlin, 1996)
7. F. Westermann, *Laser* (Teubner, 1976)
8. T. Udem, J. Reichert, R. Holzwarth, T.W. Hänsch, Opt. Lett. **24**, 881 (1999)
9. M. Wollenhaupt, A. Assion, T. Baumert, in *Springer Handbook of Lasers and Optics*, ed. by F. Träger (Springer, Berlin, 2007), chap. 12, pp. 937–983
10. D.H. Sutter, G. Steinmeyer, L. Gallmann, N. Matuschek, F. Morier-Genoud, U. Keller, V. Scheuer, G. Angelow, T. Tschudi, Opt. Lett. **24**, 631 (1999)
11. R. Trebino, *Frequency Resolved Optical Gating: the Measurement of Ultrashort Laser Pulses* (Kluwer, Boston, 2000)
12. K. Hirai, E.J. Heller, P. Gaspard, J. Chem. Phys. **103**, 5970 (1995)

2

Time-Dependent Quantum Theory

The focus of our interest will be the coupling of atomic and molecular systems to laser fields, whose maximal strength is of the order of the field exerted on an electron in the ground state of the hydrogen atom. This restriction allows us to describe the field matter interaction non-relativistically by using the time-dependent Schrödinger equation (TDSE) [1]. Analytical solutions of this linear partial differential equation are scarce, however, even in the case without external driving.

In this chapter, we continue laying the foundations for the later chapters by reviewing some basic properties of the time-dependent Schrödinger equation and by discussing two analytically solvable cases. In the following, we will then consider some general ways to rewrite, respectively, solve the time-dependent Schrödinger equation. Formulating the solution with the help of the Feynman path integral will allow us to consider an intriguing approximate, so-called semiclassical approach to the solution of the time-dependent Schrödinger equation by using classical trajectories. The last part of this chapter deals with numerical solution techniques, that will be referred to in later chapters.

2.1 The Time-Dependent Schrödinger Equation

In the heyday of quantum theory Schrödinger postulated a differential equation for the wavefunction of a quantum particle. The properties of this partial differential equation of first order in time and the interpretation of the complex-valued wavefunction are in the focus of this section. The importance of Gaussian wavepackets as (approximate) analytical solutions of the Schrödinger equation will show up for the first time by considering the so-called Gaussian wavepacket dynamics.

2.1.1 Introduction

In position representation, the time-dependent Schrödinger equation for the wavefunction of a single particle moving in three dimensions is given by

$$i\hbar\dot{\Psi}(\boldsymbol{r},t) = \hat{H}(\boldsymbol{r},t)\Psi(\boldsymbol{r},t), \tag{2.1}$$

where the Hamiltonian

$$\hat{H}(\boldsymbol{r},t) = \hat{T}_{\mathrm{k}} + V(\boldsymbol{r},t) = -\frac{\hbar^2}{2m}\Delta + V(\boldsymbol{r},t) \tag{2.2}$$

is the sum of kinetic and potential energy and the potential energy may (and in the cases to considered later will) be time-dependent.

To gain a physical interpretation of the wavefunction, one multiplies (2.1) by $\Psi^*(\boldsymbol{r},t)$ and subtracts the complex conjugate of this expression. In the case of real-valued potentials $V(\boldsymbol{r},t) = V^*(\boldsymbol{r},t)$, this procedure yields the equation of continuity

$$\dot{\rho}(\boldsymbol{r},t) = -\boldsymbol{\nabla}\cdot\boldsymbol{j}(\boldsymbol{r},t) \tag{2.3}$$

with the so-called probability density

$$\rho(\boldsymbol{r},t) = |\Psi(\boldsymbol{r},t)|^2 \tag{2.4}$$

and the probability density flux

$$\boldsymbol{j}(\boldsymbol{r},t) = \frac{\hbar}{m}\mathrm{Im}\left\{\Psi^*(\boldsymbol{r},t)\boldsymbol{\nabla}\Psi(\boldsymbol{r},t)\right\} = \frac{1}{m}\mathrm{Re}\left\{\Psi^*(\boldsymbol{r},t)\hat{\boldsymbol{p}}\Psi(\boldsymbol{r},t)\right\}. \tag{2.5}$$

We thus conclude that $|\Psi(\boldsymbol{r},t)|^2\mathrm{d}^3r$ is the probability to find a particle at time t in the volume element d^3r around $\mathbf{r}$. It may change by probability density flowing in or out, which is expressed via $\boldsymbol{j}$. Integration of (2.3) over all space yields that if Ψ is normalized at $t = t_0$ it will be normalized at all times, i.e.

$$\int_{-\infty}^{\infty} \mathrm{d}^3r|\Psi(\boldsymbol{r},t)|^2 = 1 \qquad \forall t \tag{2.6}$$

holds in the case of real potential functions and provided that the current density falls to zero faster than $1/r^2$ for $r \to \infty$.

Exercise 2.1 *Derive the equation of continuity and prove that the norm is conserved in case that $\boldsymbol{j}$ falls to zero faster than $1/r^2$ for $r \to \infty$.*

For time-independent (autonomous) potentials $V(\boldsymbol{r},t) = V(\boldsymbol{r})$ (2.1) can be solved by separation of variables. The appropriate Ansatz is

$$\Psi(\boldsymbol{r},t) = \psi(\boldsymbol{r})\varphi(t) \tag{2.7}$$

and after insertion into the time-dependent Schrödinger equation we get

$$i\hbar\frac{\dot{\varphi}(t)}{\varphi(t)} = \frac{\left\{-\frac{\hbar^2}{2m}\Delta + V(\boldsymbol{r})\right\}\psi(\boldsymbol{r})}{\psi(\boldsymbol{r})}. \tag{2.8}$$

Due to the fact that they depend on different variables, both LHS and RHS of this equation must be equal to a constant, which we name E. We thus arrive at the two equations

$$i\hbar\dot{\varphi}(t) = E\varphi(t) \tag{2.9}$$

$$\hat{H}(\boldsymbol{r})\psi_E(\boldsymbol{r}) = E\psi_E(\boldsymbol{r}). \tag{2.10}$$

The first of these equations can be solved immediately by

$$\varphi(t) = \varphi_0 e^{-iEt/\hbar}. \tag{2.11}$$

The second equation is the so-called time-independent Schrödinger equation [2]. It can be solved after specification of the potential $V(\boldsymbol{r})$, which is possible exactly analytically only in special cases, however. The energies E and the corresponding wavefunctions $\psi_E(x)$ are the eigenvalues and eigenfunctions of the problem. In their terms a particular solution of the time-dependent Schrödinger equation is given by

$$\Psi(\boldsymbol{r},t) = \psi_E(\boldsymbol{r})\varphi_0 e^{-iEt/\hbar}, \tag{2.12}$$

where the constant φ_0 later on will be absorbed in ψ_E.

Due to the linearity of the time-dependent Schrödinger equation, its most general solution is a linear combination of eigenfunctions given by

$$\Psi(\boldsymbol{r},t) = \sum_{n=0}^{\infty} a_n\psi_n(\boldsymbol{r})e^{-iE_nt/\hbar} \tag{2.13}$$

or

$$\Psi(\boldsymbol{r},t) = \int_0^{\infty} dE a(E)\psi_E(\boldsymbol{r})e^{-iEt/\hbar}, \tag{2.14}$$

where the first form is used for a discrete spectrum and the second one for the case of a continuous spectrum. Mixtures of the two cases can be dealt with by a sum of the two expressions given above.

In our derivation, we started from the time-independent Schrödinger equation to derive the time-dependent Schrödinger equation. Schrödinger, however, published them in reverse order. One can also derive the time-dependent from the time-independent version, if one considers a composite system of many degrees of freedom and traces out the so-called bath degrees of freedom leading to the "emergence of time" for the subsystem [3].

2.1.2 Time-Evolution Operator

Conservation of the norm of the wavefunction can also be proved on a more formal level by using the unitarity of the so-called time-evolution operator. This operator allows for the formal solution of the time-dependent Schrödinger equation without going to a specific representation.

For the following discussion, we therefore assume familiarity of the reader with some basic concepts of the description of quantum mechanics in Hilbert space and continue the discussion in the so-called bracket notation introduced by Dirac. Furthermore, we consider time-independent Hamiltonians to start with. The time-dependent Schrödinger equation then reads

$$i\hbar|\dot{\Psi}(t)\rangle = \hat{H}|\Psi(t)\rangle. \tag{2.15}$$

A formal solution of this equation is given by

$$|\Psi(t)\rangle = \mathrm{e}^{-i\hat{H}(t-t_0)/\hbar}|\Psi(t_0)\rangle =: \hat{U}(t,t_0)|\Psi(t_0)\rangle, \tag{2.16}$$

where we have defined the time-evolution operator $\hat{U}(t,t_0)$ which "evolves" the wavefunction from time t_0 to time t. The solution above can be easily verified by inserting it into (2.15). However, we should be careful in differentiating an exponentiated operator, see also Exercise 2.3.

With the help of the formal solution of the time-dependent Schrödinger equation it can be shown that the integral

$$\langle\Psi(t)|\Psi(t)\rangle = \langle\Psi(t_0)|\mathrm{e}^{i\hat{H}^\dagger(t-t_0)/\hbar}\mathrm{e}^{-i\hat{H}(t-t_0)/\hbar}|\Psi(t_0)\rangle = \langle\Psi(t_0)|\Psi(t_0)\rangle \tag{2.17}$$

is equal to unity for all times, if it was unity at the initial time t_0. As in the previous subsection, this is true if the Hamiltonian is Hermitian

$$\hat{H}^\dagger = \hat{H}, \tag{2.18}$$

which is equivalent to the time-evolution operator being unitary

$$\hat{U}^\dagger(t,t_0) = \hat{U}^{-1}(t,t_0), \tag{2.19}$$

as can be inferred from the definition given in (2.16). Also the composition property of the time-evolution operator

$$\hat{U}(t,t_0) = \hat{U}(t,t')\hat{U}(t',t_0) \tag{2.20}$$

can be deduced from its definition.

Things become much more involved as soon as the Hamiltonian is explicitly time-dependent. To investigate this case it is very convenient to re-express the time-dependent Schrödinger equation in terms of an integral equation. It can be shown by insertion into (2.15) that

$$|\Psi(t)\rangle = |\Psi(t_0)\rangle - \frac{i}{\hbar}\int_{t_0}^{t}\mathrm{d}t'\,\hat{H}(t')|\Psi(t')\rangle \tag{2.21}$$

is a formal solution of the time-dependent Schrödinger equation.[1] The wave-function and thus the sought for solution also appears under the integral on the RHS, however. The equation above therefore is an implicit equation. Formally, it can be solved iteratively, starting with the zeroth iteration

$$|\Psi^0(t)\rangle = |\Psi(t_0)\rangle \qquad (2.22)$$

of a constant wavefunction. The first iteration is given by

$$|\Psi^1(t)\rangle = |\Psi(t_0)\rangle - \frac{i}{\hbar} \int_{t_0}^{t} dt' \hat{H}(t')|\Psi(t_0)\rangle. \qquad (2.23)$$

After infinitely many iterations, the full solution for the time-evolution operator follows to be

$$\hat{U}(t,t_0) = \hat{1} + \sum_{n=1}^{\infty} \left(\frac{-i}{\hbar}\right)^n$$
$$\int_{t_0}^{t} dt_n \int_{t_0}^{t_n} dt_{n-1} \ldots \int_{t_0}^{t_2} dt_1 \hat{H}(t_n)\hat{H}(t_{n-1})\ldots\hat{H}(t_1), \qquad (2.24)$$

where the integration variables fulfill the inequalities $t_n \geq t_{n-1} \geq \cdots \geq t_1 \geq t_0$. One can confirm this solution by inserting it into (2.16) and finally into the time-dependent Schrödinger equation. Alternatively, one could have derived (2.24) also by "time-slicing" the interval $[t_0, t]$ into N equal parts and by successively applying the infinitesimal time-evolution operator

$$\hat{U}(t_\nu + \Delta t, t_\nu) = \exp\left\{-\frac{i}{\hbar}\hat{H}(t_\nu)\Delta t\right\} \overset{\lim \Delta t \to 0}{=} \hat{1} - \frac{i}{\hbar}\hat{H}(t_\nu)\Delta t \qquad (2.25)$$

with constant Hamiltonians at the beginning of each time step. It is rewarding to explicitly check the equivalence of this procedure with the general formula given above by working through Exercise 2.2. The fact that unitarity (and thus norm conservation) and the composition property of the time-evolution operator also hold in the time-dependent case can be most easily shown by using the decomposition in terms of infinitesimal propagators.

Exercise 2.2 *Verify the first three terms in the series for the time-evolution operator by collecting terms up to Δt^2 in the time-sliced expression*

$$\hat{U}(t,t_0) = [\hat{1} - \frac{i}{\hbar}\hat{H}(t_{N-1})\Delta t][\hat{1} - \frac{i}{\hbar}\hat{H}(t_{N-2})\Delta t]\ldots[\hat{1} - \frac{i}{\hbar}\hat{H}(t_0)\Delta t]$$

and taking the limit $N \to \infty, \Delta t \to 0$ such that $N\Delta t = t - t_0$.

---

[1] Please note that the integral form given in (2.21) is equivalent to the differential form of the time-dependent Schrödinger equation (as can be shown by differentiation) and in addition it has the initial condition "built in".

Although (2.24) gives the time-evolution operator in terms of a series, this expression is the most convenient one to work with. For reasons of completeness it shall be mentioned that a closed form expression is possible. With the definition

$$\hat{T}\left[\hat{A}(t_1)\hat{B}(t_2)\right] \equiv \begin{cases} \hat{B}(t_2)\hat{A}(t_1) & \text{if } t_2 > t_1 \\ \hat{A}(t_1)\hat{B}(t_2) & \text{if } t_1 > t_2 \end{cases}$$

of the time-ordering operator[2] it can be shown that

$$\hat{U}(t, t_0) = \hat{T}e^{-i/\hbar \int_{t_0}^{t} dt' \hat{H}(t')}. \tag{2.26}$$

Exercise 2.3 *For the verification of the formal solution of the TDSE, we need the time derivative of an operator of the form*

$$\hat{U}(t) = \exp[\hat{B}(t)].$$

(a) *Calculate $\frac{d\hat{U}}{dt}$ by using Taylor expansion of the exponential function (keep in mind that, in general, an operator does not commute with its time derivative).*
(b) *Consider the special case $\hat{B}(t) \equiv -\frac{i}{\hbar}\hat{H}_0 t$ and give a closed form solution for $\frac{d\hat{U}}{dt}$.*
(c) *Consider the special case $\hat{B}(t) \equiv -\frac{i}{\hbar}\int_0^t dt' \hat{H}(t')$ and convince yourself that a simple closed form expression for $\hat{U}(t)$ cannot be given!*
(d) *Show that the construction $\hat{U}(t) = \hat{T}\exp[\hat{B}(t)]$ with the time ordering operator and the operator $\hat{B}$ from part (c) allows for a closed form solution by proving that the relation*

$$\hat{T}\hat{B}^n = n! \left(\frac{-i}{\hbar}\right)^n \int_0^t dt_n \int_0^{t_n} dt_{n-1} \ldots \int_0^{t_2} dt_1 \hat{H}(t_n)\hat{H}(t_{n-1})\ldots\hat{H}(t_1)$$

holds.

To study some further properties of the time-evolution operator, we go into position representation again by multiplication of (2.16) from the left with $\langle r|$ and insertion of unity in terms of position states. Setting $t_0 = 0$, we find for the propagated wavefunction

$$\Psi(r, t) = \int d^3r' \langle r|\hat{U}(t, 0)|r'\rangle \Psi(r', 0). \tag{2.27}$$

The position matrix element of the time-evolution operator

$$K(r, t; r', 0) = \langle r|\hat{U}(t, 0)|r'\rangle \tag{2.28}$$

is also frequently referred to as the propagator. As can be shown by differentiation of (2.27) with respect to time, $K(r, t; r', 0)$ itself is a solution of the

---

[2] At $t_1 = t_2$ and for $\hat{A} \neq \hat{B}$ additional assumptions on ordering would have to be made.

time-dependent Schrödinger equation with initial condition $K(\boldsymbol{r}, 0; \boldsymbol{r}', 0) = \delta(\boldsymbol{r} - \boldsymbol{r}')$. For this reason, and under the assumption that $t > 0$, K is also termed *time-dependent* (retarded) Green's function of the Schrödinger equation. Again starting from (2.28) another important property of the propagator can be shown. Due to the fact that the propagator itself is a special wavefunction, the closure relation[3]

$$K(\boldsymbol{r}, t; \boldsymbol{r}', 0) = \int d^3 r'' K(\boldsymbol{r}, t; \boldsymbol{r}'', t'') K(\boldsymbol{r}'', t''; \boldsymbol{r}', 0), \qquad (2.29)$$

follows. It could have also been derived directly from (2.20) by going into position representation and inserting an additional unit operator in terms of position eigenstates.

2.1.3 Spectral Information

In the applications to be discussed in the following chapters the initial state frequently is assumed to be the ground state of the undriven problem. In this section, we will see how spectral information can, in general, be extracted form the propagator via Fourier transformation.

To extract spectral information of autonomous systems from time-series, we start from (2.28) in the case of a time-independent Hamiltonian and insert unity in terms of energy eigenstates $|\psi_n\rangle$ twice to arrive at the spectral representation

$$K(\boldsymbol{r}, t; \boldsymbol{r}', 0) = \sum_{n=0}^{\infty} \psi_n^*(\boldsymbol{r}') \psi_n(\boldsymbol{r}) \exp\left\{-\frac{i}{\hbar} E_n t\right\} \qquad (2.30)$$

of the propagator. Taking the trace (let $\boldsymbol{r}' = \boldsymbol{r}$ and integrate over $d^3 r$) of this expression one arrives at

$$\begin{aligned} G(t, 0) &= \int d^3 r\, K(\boldsymbol{r}, t; \boldsymbol{r}, 0) \\ &= \int d^3 r \sum_{n=0}^{\infty} |\psi_n(\boldsymbol{r})|^2 \exp\left\{-\frac{i}{\hbar} E_n t\right\} \\ &= \sum_{n=0}^{\infty} \exp\left\{-\frac{i}{\hbar} E_n t\right\}. \end{aligned} \qquad (2.31)$$

For large imaginary times τ only the ground-state contribution to the sum above survives. This observation leads to the so-called Feynman–Kac formula

$$E_0 = -\lim_{\tau \to \infty} \frac{1}{\tau} \ln G(-i\hbar\tau, 0). \qquad (2.32)$$

---

[3] In cases where no inverse group element exists this is sometimes also called semi-group property.

If one performs a Laplace transform on $G(t,0)$[4] then the *energy-dependent* Green's function

$$G(z) = \frac{i}{\hbar} \int_0^\infty dt G(t,0) \exp\left\{\frac{i}{\hbar} zt\right\}$$

$$= \sum_{n=0}^\infty \frac{1}{E_n - z} \tag{2.33}$$

emerges. This function has poles at the energy levels of the underlying eigenvalue problem.

For numerical purposes it is often helpful to study the time-evolution of wavepackets. By considering the auto-correlation function of an initial wavefunction

$$|\Psi_\alpha\rangle = \sum_n |n\rangle\langle n|\Psi_\alpha\rangle = \sum_n c_n|n\rangle, \tag{2.34}$$

which is defined according to

$$c_{\alpha\alpha}(t) = \langle \Psi_\alpha | e^{-i\hat{H}t/\hbar} | \Psi_\alpha\rangle = \sum_n |c_n|^2 e^{-iE_n t/\hbar}, \tag{2.35}$$

one gains the local spectrum by Fourier transformation

$$S(\omega) = \frac{1}{2\pi\hbar} \int dt e^{i\omega t} c_{\alpha\alpha}(t)$$

$$= \sum_{n=0}^\infty |c_n|^2 \delta(E_n - \hbar\omega). \tag{2.36}$$

This result is a series of peaks at the eigenvalues of the problem, that are weighted with the absolute square of the overlap of the initial state with the corresponding eigenstate $|n\rangle$. A recent development in this area is the use of so-called harmonic inversion techniques instead of Fourier transformation. This is a nonlinear procedure which allows the use of rather short time series to extract spectral information [4].

Not only the spectrum but also the eigenfunctions can be determined from time-series. To this end one considers the Fourier transform of the wavefunction at one of the energies just determined

$$\lim_{T\to\infty} \frac{1}{2T} \int_{-T}^T dt e^{iE_m t/\hbar} |\Psi_\alpha(t)\rangle = \sum_{n=0}^\infty c_n \lim_{T\to\infty} \frac{1}{2T} \int_{-T}^T dt e^{-i(E_n - E_m)t/\hbar} |n\rangle$$

$$= c_m|m\rangle. \tag{2.37}$$

---

[4] To ensure convergence of the integral, one adds a small positive imaginary part to the energy, $z = E + i\epsilon$.

This procedure filters an eigenfunction if the overlap with the initial state is sufficiently high. Alternatively, the concept of imaginary time propagation is again helpful if one wants to determine e.g., the ground state. To this end, the time evolution of (2.34)

$$|\Psi_\alpha(-i\hbar\tau)\rangle = \sum_n c_n |n\rangle e^{-\tau E_n} \tag{2.38}$$

is considered for large τ, when only the ground state contribution survives. At the end of the calculation, the ground state has to be renormalized and can be subtracted from the initial wavefunction. Repeating the procedure with the modified initial state, the next highest energy state can in principle be determined.

2.1.4 Analytical Solutions for Wavepackets

In the following, we will consider an Ansatz for the solution of the time-dependent Schrödinger equation with the help of a Gaussian wavepacket. For potentials that are maximally quadratic, this approach leads to an exact analytic solution. It is also applicable to driven problems and to nonlinear forces. In the last case, the method is approximate in nature, however.

To finish this introductory section, we then review the dynamics of a wavepacket in the box potential.

Gaussian Wavepacket Dynamics

As early as 1926, Schrödinger has stressed the central importance of Gaussian wavepackets in the transition from "micro-" to "macro-mechanics" [5]. For this reason we will now consider what happens if we make a Gaussian Ansatz for the solution of the time-dependent Schrödinger equation.

In Heller's so-called Gaussian Wavepacket Dynamics (GWD) [6], a complex-valued Gaussian (here for reasons of simplicity in one spatial dimension)

$$\Psi(x,t) = \left(\frac{2\alpha_0}{\pi}\right)^{1/4} \exp\left\{-\alpha_t(x-q_t)^2 + \frac{i}{\hbar}p_t(x-q_t) + \frac{i}{\hbar}\delta_t\right\} \tag{2.39}$$

is used as an Ansatz for the solution of (2.1). The expression above contains *real valued* parameters $q_t, p_t, \alpha_0 = \alpha_{t=0} \in \mathbf{R}$ and *complex valued* ones $\alpha_t, \delta_t \in \mathbf{C}$. The initial width parameter α_0 shall be given. The parameters $q_t, p_t, \alpha_t, \delta_t$ that are undetermined up to now can be gained by a Taylor expansion of the potential around $x = q_t$ according to

$$V(x,t) \approx V(q_t,t) + V'(q_t,t)(x-q_t) + \frac{1}{2!}V''(q_t,t)(x-q_t)^2 \tag{2.40}$$

and using the time-dependent Schrödinger equation. After insertion of the time and position derivatives of the wavefunction

$$\dot{\Psi}(x,t) = \left\{ -\dot{\alpha}_t(x-q_t)^2 + 2\alpha_t\dot{q}_t(x-q_t) + \frac{i}{\hbar}\dot{p}_t(x-q_t) - \frac{i}{\hbar}p_t\dot{q}_t + \frac{i}{\hbar}\dot{\delta}_t \right\}$$
$$\Psi(x,t)$$

$$\Psi'(x,t) = \left[-2\alpha_t(x-q_t) + \frac{i}{\hbar}p_t \right]\Psi(x,t)$$

$$\Psi''(x,t) = \left\{ -2\alpha_t + \left[-2\alpha_t(x-q_t) + \frac{i}{\hbar}p_t \right]^2 \right\}\Psi(x,t)$$

the coefficients of equal powers of $x - q_t$ can be compared. This leads to the following set of equations

(i) $(x-q_t)^2$: $-i\hbar\dot{\alpha}_t = -\dfrac{2\hbar^2}{m}\alpha_t^2 + \dfrac{1}{2}V''(q_t,t),$

(ii) $(x-q_t)^1$: $i\hbar2\alpha_t\dot{q}_t - \dot{p}_t = i\hbar2\alpha_t\dfrac{p_t}{m} + V'(q_t,t),$

(iii) $(x-q_t)^0$: $p_t\dot{q}_t - \dot{\delta}_t = \dfrac{\hbar^2}{m}\alpha_t + \dfrac{p_t^2}{2m} + V(q_t,t).$

From (ii), after separation of real and imaginary part, Hamilton equations for q_t and p_t emerge

$$\dot{q}_t = \frac{p_t}{m}, \tag{2.41}$$

$$\dot{p}_t = -V'(q_t,t). \tag{2.42}$$

From (iii) and with the definition of the classical Lagrangian

$$L = T_k - V \tag{2.43}$$

the equation

$$\dot{\delta}_t = L - \frac{\hbar^2}{m}\alpha_t \tag{2.44}$$

can be derived. The remaining equation for the width parameter α_t follows from (i). It is the nonlinear Riccati differential equation

$$\dot{\alpha}_t = -\frac{2i\hbar}{m}\alpha_t^2 + \frac{i}{2\hbar}V''(q_t,t). \tag{2.45}$$

The width of the Gaussian is a function of time and in contrast to an approach, that will be discussed later in this chapter, where the width of the Gaussians is fixed (i.e., "frozen"), the approach reviewed here is also referred to as the "thawed" GWD.

In the cases of the free particle and of the harmonic oscillator, the procedure above leads to an exact analytic solution of the time-dependent Schrödinger equation.

Exercise 2.4 *Use the GWD-Ansatz to solve the TDSE.*

(a) Use the differential equations for $q_t, p_t, \alpha_t, \delta_t$ to show that the Gaussian wavepacket fulfills the equation of continuity.

(b) Solve the differential equations for $q_t, p_t, \alpha_t, \delta_t$ for the free particle case $V(x) = 0$.

(c) Solve the differential equations for $q_t, p_t, \alpha_t, \delta_t$ for the harmonic oscillator case $V(x) = \frac{1}{2} m\omega_e^2 x^2$.

(d) Calculate the expectation values and variances $\langle x \rangle$, $\langle p \rangle$, $\Delta x = \sqrt{\langle x^2 \rangle - \langle x \rangle^2}$, $\Delta p = \sqrt{\langle p^2 \rangle - \langle p \rangle^2}$ and from this the uncertainty product using the general Gaussian wavepacket. Discuss the special cases from (b) and (c). What results do you get for the harmonic oscillator in the case $\alpha_{t=0} = \frac{m\omega_e}{2\hbar}$?

The GWD method can also be applied to nonlinear classical problems, however, where it is typically valid only for short times. In this context, one often uses the notion of the Ehrenfest time, after which non-Gaussian distortions of a wavepacket become manifest.

Particle in an Infinitely High Square Well

As a final example for an exactly solvable problem in quantum dynamics, we now consider the evolution of an initial wavepacket in an infinitely high square well (box potential). This problem has been presented in the December 1995 issue of "Physikalische Blätter" [7] and leads to aesthetically pleasing space–time pictures, which are sometimes referred to as "quantum carpets".

In the work of Kinzel, a Gaussian initial state localized close to the left wall has been investigated. For ease of analytic calculations let us consider a wavefunction made up of a sum of eigenfunctions with equal weight in the following. The eigenfunctions and eigenvalues of the square well extending from 0 to L are given by

$$\psi_n(x) = \sqrt{\frac{2}{L}} \sin\left\{ \frac{n\pi}{L} x \right\} \tag{2.46}$$

$$E_n = n^2 E_1, \qquad n = 1, 2, 3, \dots \tag{2.47}$$

with the fundamental energy

$$E_1 = \frac{1}{2m} \left(\frac{\hbar\pi}{L} \right)^2. \tag{2.48}$$

The corresponding frequency and period are

$$\omega = E_1/\hbar; \qquad T = \frac{2\pi}{\omega}.$$

Exercise 2.5 *A particle is in the eigenstate ψ_n with energy E_n of an infinitely high potential well of width L ($0 \leq x \leq L$). Let us assume that the width of the well is suddenly doubled.*

(a) Calculate the probability to find a particle in the eigenstate ψ'_m with energy E'_m of the new well.

(b) Calculate the probability to find a particle in state ψ'_m whose energy E'_m is equal to E_n.

(c) Consider the time evolution for $n = 1$, i.e., at $t = 0$ the wavefunction is the lowest eigenfunction of the small well $\Psi(x, 0) = \psi_1(x)$. Calculate the smallest time $t_{\min}$ for which $\Psi(x, t_{\min}) = \Psi(x, 0)$.

(d) Draw a picture of the wavefunction $\Psi(x, t)$ at $t = t_{\min}/2$.

In the following, we will use dimensionless variables for position and time

$$\xi = \frac{x}{L}; \qquad \tau = \frac{t}{T}$$

and will consider an initial wavefunction consisting of N eigenfunctions with equal weights. As can be seen in figure 1 of [8] for large N, the wavefunctions have the same initial localization properties as the ones used in [7]. Each eigenfunction is evolving with the exponential of the corresponding eigenenergy according to (2.12)). The time-evolution of the normalized wavepacket thus is

$$\Psi(\xi, \tau) = \sqrt{\frac{2}{LN}} \sum_{n=1}^{N} \sin(n\pi\xi) \exp(2\pi i n^2 \tau). \qquad (2.49)$$

In Fig. 2.1, we show the absolute value of the time-evolved wavefunction with $N = 20$ in the range $0 \le \xi \le 1; 0 \le \tau \le 0.5$ of the $\xi - \tau$ plane. In this density plot, darkness corresponds to a low and brightness to a large value of the plotted function. A detailed explanation of the features of the time-evolved wavefunction has been given in [8]. Let us here concentrate on the suppression of the wavefunction along the lines $\tau = \frac{\xi}{2k}$. To make progress, we express the sine function with the help of exponentials and arrive at terms of the form

$$\vartheta(\pm\xi, \tau) = \sum_{n=1}^{N} q^{n^2} e^{\pm i n \pi \xi} \qquad (2.50)$$

with $q \equiv e^{2\pi i \tau}$. We now shift the argument ξ by $2k\tau$, with positive $k = 1, 2, 3, \ldots$. Using $n = n + k/2 - k/2$ we get

$$\vartheta(\xi + 2k\tau, \tau) = \sum_{n=1}^{N} q^{n^2} e^{i n \pi (\xi + 2k\tau)}$$

$$= q^{-(k/2)^2} e^{-i(k/2)\pi\xi} \sum_{n=1}^{N} q^{(n+k/2)^2} e^{i(n+k/2)\pi\xi}$$

$$\vartheta(-(\xi + 2k\tau), \tau) = \sum_{n=1}^{N} q^{n^2} e^{-i n \pi (\xi + 2k\tau)}$$

$$= q^{-(k/2)^2} e^{-i(k/2)\pi\xi} \sum_{n=1}^{N} q^{(n-k/2)^2} e^{-i(n-k/2)\pi\xi}.$$

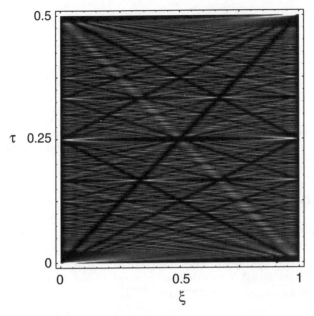

Fig. 2.1. Time-evolution of a superposition of 20 eigenfunctions of the box potential [8]

The wavefunction along the straight lines originating from $\xi = 0$ can thus be written as

$$\Psi(2k\tau, \tau) = \sqrt{\frac{2}{LN}} \frac{1}{2i} q^{-(k/2)^2} \sum_{n=1}^{N} \{q^{(n+k/2)^2} - q^{(n-k/2)^2}\}$$

$$= \sqrt{\frac{2}{LN}} \frac{1}{2i} q^{-(k/2)^2}$$

$$\{q^{(N+k/2)^2} + \ldots + q^{(N+1-k/2)^2} - q^{(k/2)^2} - \ldots - q^{(1-k/2)^2}\}.$$

Using the analogy with two combs shifted against each other, as shown in Fig. 2.2 it is obvious that from the expression in curly brackets above, only $2k$ terms of order 1 do survive, due to the fact that the major part of the sum cancels term-wise. In the case $k \ll \sqrt{N}$, we thus can conclude

$$\Psi(2k\tau, \tau) \sim k/\sqrt{N} \approx 0$$

These considerations thus explain the "channels" of near extinction of the wavefunction along lines of slope $1/2k$ emanating from $(\xi = 0, \tau = 0)$ to the right. For a given value of N, the higher the value of k fewer terms cancel each other and the less visible the "channel effect" becomes.

Fig. 2.2. Comb analogy for $N = 20$ and $k = 2$. The overlapping parts of the two combs are representing the terms that cancel each other. Only the terms without partner (here 4) do not cancel

2.2 Analytical Approaches

For most problems of interest a direct exact analytical solution of the time-dependent Schrödinger equation cannot be found. It is therefore of quite some interest to devise alternative approaches to quantum dynamics and/or some approximate or exact Ansätze that are generally applicable and lead to viable approximate and/or numerical schemes.

One reformulation of the time-dependent Schrödinger equation e.g., is the Feynman path integral expression for the propagator [9]. This is of utmost importance in the following because from the path integral a much used approximation can be derived: the time-dependent semiclassical formulation of quantum theory. Furthermore, in the case of small external perturbations, time-dependent perturbation theory may be the method of choice for the solution of the time-dependent Schrödinger equation. Moreover, for systems with many degrees of freedom, as a first approximation the wavefunction can be factorized. We thus discuss the so-called Hartree Ansatz, and for the first time also the Born–Oppenheimer method in this chapter. Finally, an exact analytical Ansatz for the solution of periodically driven quantum systems is given by Floquet theory.

The discussion of the numerical implementation of some of these concepts will be postponed to Sect. 2.3.

2.2.1 Feynman's Path Integral

For time-dependent quantum problems, which occur naturally if we want to describe the interaction of a system with a laser field, as we will see in Chap. 3, an approach that deals with the propagator is very well suited. With the propagator at hand, we can calculate the time evolution of every wavefunction according to (2.27).

The Propagator as a Path Integral

A very elegant approach, that gives an explicit formula for the propagator goes back to an idea that can e.g., be found in later editions of the famous text book

by Dirac [10] and has finally been formulated by Feynman [9]. The derivation
of the path integral is a prime example for the new quantum mechanical
reasoning in terms of probability amplitudes in contrast to the classical way
of thinking in probabilities. The famous double slit experiment serves as the
chief parable to understand the new twist of quantum mechanical thinking.
The postulates that form the basis for the derivation of the path integral are:

Postulate 1: If, in the particle picture, an event can have occurred in two
  mutually exclusive ways, the corresponding amplitudes have to be added
  to find the total amplitude
Postulate 2: If an event consists of two successive events, the corresponding
  amplitudes do multiply.

In the book by Feynman and Hibbs [11], it is shown how the usage of the two
axioms given above leads to the composition (or semigroup) property of the
propagator, which we have already stated in (2.29). At this point, however,
only physical intuition helps. How can the probability amplitude $K(\boldsymbol{r}_f, t; \boldsymbol{r}_i, 0)$
itself be determined?

Inspired by an idea of Dirac, and using the first postulate above, Feynman
in 1948 [9] expressed the propagator as a sum over all paths from $\boldsymbol{r}_i$ to $\boldsymbol{r}_f$
in time t. Each of these paths contributes with a phase factor to the sum.
The phase is the ratio of the classical action of the respective path and $\hbar$.
Mathematically this can be written as

$$K(\boldsymbol{r}_f, t; \boldsymbol{r}_i, 0) = \int_{\boldsymbol{r}(0)=\boldsymbol{r}_i}^{\boldsymbol{r}(t)=\boldsymbol{r}_f} \mathrm{d}[\boldsymbol{r}] \exp\left\{ \frac{\mathrm{i}}{\hbar} S[\boldsymbol{r}] \right\} \tag{2.51}$$

with Hamilton's principal function (being a functional of the path, which is
expressed by the square brackets)

$$S[\boldsymbol{r}] = \int_0^t \mathrm{d}t'\, L. \tag{2.52}$$

The classical Lagrangian has already appeared in (2.43). The symbol $\int \mathrm{d}[\boldsymbol{r}]$
is denoting the integral over all paths (functional integral). In contrast to
standard integration, where one sums a function over a certain range of a
variable, in a path integral one sums a function of a function (a so-called
functional) over a certain class of *functions* that are parametrized by t and
that obey the boundary conditions $\boldsymbol{r}(0) = \boldsymbol{r}_i$, $\boldsymbol{r}(t) = \boldsymbol{r}_f$. Feynman's path
integral is therefore sometimes also referred to as "sum over histories". In
Fig. 2.3, for an arbitrary one-dimensional potential, we depict the classical
path and some other equally important nonclassical paths.

The exact analytic calculation of the path integral, apart from a few ex-
ceptions involving quadratic Lagrangians, is not possible. One therefore frequently resorts to approximate solutions. A principal way to calculate the
path integral shall, however, be hinted at. To make progress, the time interval

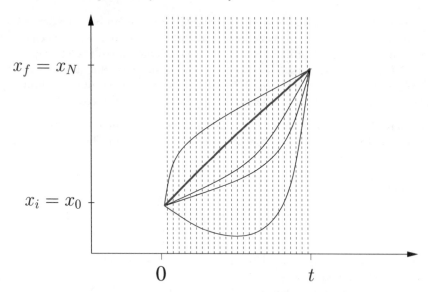

Fig. 2.3. Paths is space–time. Time-slicing and the classical path (*thick grey line*) are also depicted

$[0, t]$ is divided into N equal parts of length Δt, analogously to the procedure in Sect. 2.1.2. This "time-slicing" is depicted in Fig. 2.3. In this way the path integral is discretized and in the limit $N \to \infty$ can be written as an $(N-1)$-dimensional Riemann integral times a normalization constant B_N. In one spatial dimension this reads

$$K(x_f, t; x_i, 0) = \lim_{N \to \infty} B_N \int dx_1 \int \dots \int dx_{N-1}$$
$$\exp\left\{ \frac{i}{\hbar} \sum_{j=1}^{N} \left[\frac{(x_j - x_{j-1})^2}{2\Delta t} - V\left(\frac{x_j + x_{j-1}}{2} \right) \Delta t \right] \right\}. \quad (2.53)$$

Proving this expression is most elegantly done by using a Weyl transformation and will be performed explicitly in Appendix 2.A. Furthermore, the expression above can be interpreted as the successive application of the closure relation concatenating short-time propagators

$$K(x_j, \Delta t; x_{j-1}, 0) \sim \exp\left\{ \frac{i}{\hbar} \left[\frac{(x_j - x_{j-1})^2}{2\Delta t} - V\left(\frac{x_j + x_{j-1}}{2} \right) \Delta t \right] \right\}. \quad (2.54)$$

It considerably deepens one's understanding to derive the short-time propagator directly from the infinitesimal time-evolution operator of (2.25). The interesting question how the Hamilton operator "mutates" into the classical Lagrangian is answered in Exercise 2.6. In the second part of this exercise, the time-dependent Schrödinger equation can be derived. To this end a simplified version of the short-time propagator with a simple end point rule for the discretization of the potential part of the action by replacing

$$V\left(\frac{x_j + x_{j-1}}{2}\right) \qquad \text{by} \qquad V(x_{j-1}) \qquad (2.55)$$

in (2.54) is sufficient.

Exercise 2.6

(a) Derive the short-time propagator starting from

$$\hat{U}(\Delta t) = \exp\{-i\hat{H}\Delta t/\hbar\}$$

for the infinitesimal time-evolution operator.
Hint: Use first-order Taylor expansion of the exponential function
(b) Use the short-time propagator to propagate an arbitrary wavefunction
$\Psi(x, t)$ over an infinitesimal time interval Δt via

$$\Psi(x, t + \Delta t) = \int dy K(x, \Delta t; y, 0) \Psi(y, t).$$

and derive the TDSE!
Hint: To this integral only a small interval of y centered around x is contributing. Expansion of the expression above to first order in Δt and up to second order in $\eta = y - x$ leads to a linear differential equation for $\Psi(x, t)$. Use Gaussian integrals, where appropriate.

Stationary Phase Approximation

By inspection of (2.53) it is obvious, that the calculation of the propagator for arbitrary potentials becomes arbitrarily complicated. In the case of maximally quadratic potentials all integrals are Gaussian integrals[5], however, and thus can be done exactly analytically. There are some additional examples, for which exact analytic results for the path integral are known. These are collected in the supplement section of the Dover edition of the textbook by Schulman [12].

In general, however, approximate solutions for the path integral are sought for. The idea is to approximate the exponent in such a fashion that only quadratic terms survive. The corresponding approximation is the stationary phase approximation (SPA). It shall be introduced by first looking at a simple 1d integral of the form

$$\int_{-\infty}^{+\infty} dx \, \exp\{if(x)/\delta\}g(x). \qquad (2.56)$$

---

[5] For purely imaginary exponent sometimes also referred to as Fresnel integrals.

To proceed, we perform a Taylor expansion of the function in the exponent up to second order according to $f(x) \approx f(x_0)+f'(x_0)(x-x_0)+1/2f''(x_0)(x-x_0)^2$ under the condition of stationarity of the phase, i.e.,

$$f'(x_0) = 0. \tag{2.57}$$

Then with the help of the formula (Fresnel-Integral, i.e., (1.28) with purely imaginary a)

$$\int_{-\infty}^{+\infty} dx \exp(i\alpha x^2) = \sqrt{\frac{i\pi}{\alpha}}, \tag{2.58}$$

we get

$$\int_{-\infty}^{+\infty} dx \exp\{if(x)/\delta\}g(x) \stackrel{\delta \to 0}{=} \sqrt{\frac{2\pi i\delta}{f''(x_0)}} \exp\{if(x_0)/\delta\}g(x_0) \tag{2.59}$$

for the integral above. This approximation becomes the better, the faster the exponent oscillates, which is determined by the smallness of the parameter δ. To demonstrate this fact, in Fig. 2.4 the function $\exp\{ix^2/\delta\}$ is displayed using $\delta = 0.01$. The fast oscillations of the function at x values further away than $\sqrt{\delta}$ from the point of stationary phase, lead to the mutual cancellation of the positive and negative contributions to the integral. Around the stationary phase point (which here is $x = 0$) this argument does not apply and therefore the major contribution to the integral is determined by the properties of the function $f(x)$ around that point.

In Sect. 2.2.2, the notion of stationary phase integration will be extended to the path integral, being an infinite dimensional "normal" integral. Before

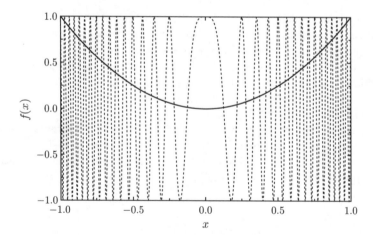

Fig. 2.4. The function $f(x) = x^2$ (*solid line*) and the real part of $\exp\{ix^2/\delta\}$ with $\delta = 0.01$ (*dashed line*)

doing so, a remark on the direct numerical approach to the path integral is in order. As can be seen by looking at the integrand of our 1d toy problem, a numerical attack to calculate the integral of a highly oscillatory function will be problematic due to the near cancellation of terms. This is even more true for the full-fledged path integral and the associated problem is sometimes referred to as the *sign problem*, which is a topic at the forefront of present-day research. Much more well-behaved with respect to numerical treatment are imaginary time path integrals, which will not be dealt with herein, however.

2.2.2 Semiclassical Approximation

The semiclassical approximation of the propagator goes back to van Vleck [13]. Its direct derivation from the path integral followed many years later, however, and finally led to the generalization of the van Vleck formula by Gutzwiller [14]. We will later-on use semiclassical arguments quite frequently, because they allow for a qualitative and often also for a quantitative explanation of many interesting quantum phenomena. For this reason, we will go through the derivation of the so-called van Vleck–Gutzwiller (VVG) propagator in some detail.

To derive a time-dependent semiclassical expression, the SPA will be applied to the path integral (2.51). In this case, the exponent $S[x]$ is a functional. Therefore, we need the definition of the variation of a functional (see e.g., [15]). The analog of (2.57) is

$$\delta S[x_{cl}] = 0. \tag{2.60}$$

This, however, is Hamilton's principle of classical mechanics. The SPA is thus based on the expansion of the exponent of the path integral around the classical path. This is also the reason why we have highlighted the classical path in Fig. 2.3. By defining the deviation from the classical path as

$$\eta(t') = x(t') - x_{cl}(t'), \qquad \eta(0) = \eta(t) = 0, \tag{2.61}$$

the second-order expansion needed for the SPA is given by

$$S[x] = S[x_{cl}] + \frac{1}{2} \int_0^t dt' \, \eta(t') \hat{O} \eta(t') \tag{2.62}$$

with

$$\hat{O} = -m \frac{d^2}{dt^2} - V'' \big|_{x = x_{cl}(t)}. \tag{2.63}$$

More details about the underlying variational calculus can be found in Chap. 12 of [12] and in Appendix 2.B.

For the time being, we assume that only one single point of stationary phase exists. In SPA the propagator can then be written as

$$K(x_f, t; x_i, 0) \approx \int_{\eta(0)=0}^{\eta(t)=0} d[\eta] \exp\left\{\frac{i}{2\hbar} \int_0^t dt' \eta(t') \hat{O} \eta(t')\right\},$$

$$\exp\left\{\frac{i}{\hbar} S[x_{cl}]\right\}. \tag{2.64}$$

Due to the additive nature of the action in (2.62) the propagator factorizes into a trivial factor coming from the zeroth-order term in the expansion and a so-called prefactor coming from the fluctuations around the classical path. Due to its boundary conditions, this prefactor is also referred to as 0–0-propagator.

The condition for the applicability of the SPA is that the exponent oscillates rapidly. For the functional integral, this means that $\hbar$ must be small compared to the action of the classical trajectory. The main contribution to the propagator in the SPA thus stems from the classical path that solves the boundary value problem defined by the propagator labels and from a small region surrounding the classical path. In this context, the SPA is therefore also called the semiclassical approximation. There are several ways to derive an explicit expression for the prefactor in (2.64), see e.g., [12]. As shown in detail in this textbook, the final expression for the propagator is given by

$$K(x_f, t; x_i, 0) \approx \sqrt{\frac{i}{2\pi\hbar} \frac{\partial^2 S[x_{cl}]}{\partial x_f \partial x_i}} \exp\left\{\frac{i}{\hbar} S[x_{cl}]\right\}. \tag{2.65}$$

The classical information that enters the expression above can be gained by solving the root search problem defined by the propagator labels and calculating the corresponding action and its second derivative with respect to the initial and final position.

As mentioned at the beginning of this section, van Vleck succeeded already in 1928 in finding the above expression in a more "heuristic" manner [13]. He started his derivation from the observation that the insertion of the Ansatz[6]

$$K \sim \exp\{iS(t)/\hbar\} \tag{2.66}$$

into the time-dependent Schrödinger equation in the limit of $\hbar \to 0$ leads to the Hamilton–Jacobi equation

$$H\left(\frac{\partial S}{\partial x(t)}, x(t)\right) + \frac{\partial S}{\partial t} = 0. \tag{2.67}$$

Invoking the correspondence principle, S must thus be the classical action functional, and we found the exponential part of the propagator. The prefactor

---

[6] This Ansatz is also the starting point of so-called Bohmian methods to study quantum dynamics. For a very good discussion of this topic, see e.g., Chap. 4 of [16].

follows from a more involved reasoning: the classical density for reaching point x_f after starting from point x_i can be determined by integrating over all possible initial momenta (vertical line in the phase space plot in Fig. 2.5) and is given by

$$\rho_{\text{cl}}(x_f, t; x_i, 0) = \int dp_i \delta[x_f - x_t(x_i, p_i)] = \left| \frac{\partial x_f}{\partial p_i} \Big|_{x_i} \right|^{-1}, \qquad (2.68)$$

where for the time being, we have again assumed that there is just one single solution of the double-ended boundary value problem. Converting to quantum mechanical amplitudes, after dividing by h, the square root of the classical density has to be taken to arrive at the correct semiclassical prefactor. From basic classical mechanics we have the identity

$$\frac{\partial^2 S[x_{\text{cl}}]}{\partial x_f \partial x_i} = -\frac{\partial p_i}{\partial x_f}, \qquad (2.69)$$

however. Thus, up to the absolute value of the radicant, which is due to the density being a positive definite quantity, the semiclassical propagator of van Vleck and the one from the SPA to the path integral are equivalent.

Almost 40 years later, Gutzwiller has extended the validity of the van Vleck expression to longer times [14]. First of all, for finite times there may be several solutions to the classical root search problem. Such a situation is depicted graphically in Fig. 2.5. In the case of multiple solutions an additional summation therefore has to be introduced in (2.65). Using the path integral derivation together with the summation over several points of stationary

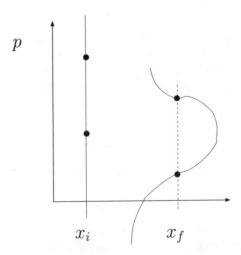

Fig. 2.5. The case of two classical solutions of the boundary value problem. The phase space manifold of initial conditions is indicated by the solid vertical line. This manifold evolves into the s-shaped curve, whereby two of the initial conditions, indicated by the *dots*, fulfill the boundary value problem

phase, one finally arrives at the van Vleck–Gutzwiller expression

$$K^{\text{VVG}}(x_f, t; x_i, 0) \equiv \sum_{\text{cl}} \sqrt{\frac{\text{i}}{2\pi\hbar} \left| \frac{\partial^2 S[x_{\text{cl}}]}{\partial x_f \partial x_i} \right|} \exp\left\{ \frac{\text{i}}{\hbar} S[x_{\text{cl}}] - \text{i}\nu\,\pi/2 \right\} \quad (2.70)$$

with the so-called Maslov (or Morse) index ν, introduced into the semiclassical propagator by Gutzwiller and counting the caustics a path has gone through. The Maslov phase allows one to use the absolute value inside the square root.

This final expression has interference effects built in, because of the summation over classical trajectories and is very elegant and intuitive, because it relies solely on classical trajectories. However, it also has a major drawback, especially for systems with several degrees of freedom. Then, the underlying root search problem becomes extremely hard to solve and a semiclassical propagator using the solution of classical *initial value problems* would be much needed. Such a reformulation of the semiclassical expression is possible and will be discussed in Sect. 2.3.4 on numerical methods.

2.2.3 Time-Dependent Perturbation Theory

Another approximate way to solve the time-dependent Schrödinger equation starts directly from the infinite sum of time-ordered operator products in (2.24) for the time-evolution operator. Considering this series as a perturbation expansion and taking only a few terms into account will lead to reasonable results only for short times. In the case of additive Hamiltonians

$$\hat{H}(t) = \hat{H}_0(t) + \hat{W}(t) \quad (2.71)$$

we can, however, split the problem into parts and can possibly treat the first one analytically and the rest perturbatively. Please note that in (2.71) both the first and second terms may depend on time. This will come in handy when we discuss the different pictures of quantum mechanics in a unified manner.

The time-evolution operator for the unperturbed problem is formally given by

$$\hat{U}_0(t, 0) = \hat{T} \exp\left\{ -\frac{\text{i}}{\hbar} \int_0^t \text{d}t'\, \hat{H}_0(t') \right\}. \quad (2.72)$$

With its help, we define a wavefunction in the interaction picture (indicated by the index I)

$$|\Psi_{\text{S}}(t)\rangle =: \hat{U}_0(t, 0)|\Psi_{\text{I}}(t)\rangle, \quad (2.73)$$

where the wavefunction $|\Psi_{\text{S}}(t)\rangle$ is the one in the Schrödinger picture which we have considered up to now. Inserting (2.73) into the time-dependent Schrödinger equation and using

$$\text{i}\hbar\dot{\hat{U}}_0(t, 0) = \hat{H}_0(t)\hat{U}_0(t, 0) \quad (2.74)$$

yields

$$\text{i}\hbar|\dot{\Psi}_{\text{I}}(t)\rangle = \hat{W}_{\text{I}}(t)|\Psi_{\text{I}}(t)\rangle, \quad (2.75)$$

i.e., the Schrödinger equation in the interaction picture, where the perturbation Hamiltonian in the interaction picture is given by

$$\hat{W}_I(t) := \hat{U}_0^\dagger(t,0)\hat{W}(t)\hat{U}_0(t,0). \tag{2.76}$$

The time-evolution operator in the interaction picture is

$$\hat{U}_I(t,0) := \hat{T}\exp\left\{-\frac{i}{\hbar}\int_0^t dt'\hat{W}_I(t')\right\}. \tag{2.77}$$

Using (2.73), we note that at $t = 0$ the wavefunctions are identical, i.e., $|\Psi_S(0)\rangle = |\Psi_I(0)\rangle = |\Psi(0)\rangle$.

Furthermore, calculating an expectation value in the Schrödinger picture and using (2.73), we get

$$\begin{aligned}\langle\hat{A}\rangle(t) &= \langle\Psi_S(t)|\hat{A}_S|\Psi_S(t)\rangle\\ &= \langle\Psi_I(t)|\hat{U}_0^\dagger(t,0)\hat{A}_S\hat{U}_0(t,0)|\Psi_I(t)\rangle\end{aligned} \tag{2.78}$$

With the definition of an operator in the interaction picture

$$\hat{A}_I(t) := \hat{U}_0^\dagger(t,0)\hat{A}_S\hat{U}_0(t,0), \tag{2.79}$$

the expectation value becomes

$$\langle\hat{A}\rangle(t) = \langle\Psi_I(t)|\hat{A}_I(t)|\Psi_I(t)\rangle, \tag{2.80}$$

which is identical to the Schrödinger picture expectation value.

Exercise 2.7 *Verify that the time-evolution operator in the interaction picture* $\hat{U}_I(t,0) = \hat{U}_0^\dagger(t,0)\hat{U}(t,0)$ *fulfills the appropriate differential equation.*

The third picture that is frequently applied is the Heisenberg picture . By an appropriate choice of the unperturbed Hamiltonian and the perturbation all three pictures can be dealt with in the same framework:

- $\hat{H} = \hat{H}_0 + \hat{W}$ leads to the interaction picture
- $\hat{H}_0 = 0$ and $\hat{W} = \hat{H}$ leads to the Schrödinger picture
- $\hat{H}_0 = \hat{H}$ and $\hat{W} = 0$ leads to the Heisenberg picture

The relations between the different cases are given in Table 2.1 for the wavefunctions and in Table 2.2 for the operators.

With the interaction picture defined, we can now derive time-dependent perturbation theory for small perturbations $\hat{W}(t)$. Iterative solution of the corresponding time-dependent Schrödinger equation leads to an expression for the propagator in the interaction picture

$$\hat{U}_I(t,0) = \hat{1} + \sum_{n=1}^\infty \left(\frac{-i}{\hbar}\right)^n$$

$$\int_0^t dt_n \int_0^{t_n} dt_{n-1}\dots\int_0^{t_2} dt_1\hat{W}_I(t_n)\hat{W}_I(t_{n-1})\dots\hat{W}_I(t_1), \tag{2.81}$$

Table 2.1. Relations between the wavefunctions in the different pictures of quantum mechanics

	$|\Psi_S(t)\rangle$	$|\Psi_H\rangle$	$|\Psi_I(t)\rangle$
$|\Psi_S(t)\rangle$		$\hat{U}(t,0)|\Psi_H\rangle$	$\hat{U}_0(t,0)|\Psi_I(t)\rangle$
$|\Psi_H\rangle$	$\hat{U}^\dagger(t,0)|\Psi_S(t)\rangle$		$\hat{U}_I^\dagger(t,0)|\Psi_I(t)\rangle$
$|\Psi_I(t)\rangle$	$\hat{U}_0^\dagger(t,0)|\Psi_S(t)\rangle$	$\hat{U}_I(t,0)|\Psi_H\rangle$	

Table 2.2. Relations between the operators in the different pictures of quantum mechanics

	$\hat{A}_S$	$\hat{A}_H(t)$	$\hat{A}_I(t)$
$\hat{A}_S$		$\hat{U}(t,0)\hat{A}_H(t)\hat{U}^\dagger(t,0)$	$\hat{U}_0(t,0)\hat{A}_I(t)\hat{U}_0^\dagger(t,0)$
$\hat{A}_H(t)$	$\hat{U}^\dagger(t,0)\hat{A}_S\hat{U}(t,0)$		$\hat{U}_I^\dagger(t,0)\hat{A}_I(t)\hat{U}_I(t,0)$
$\hat{A}_I(t)$	$\hat{U}_0^\dagger(t,0)\hat{A}_S\hat{U}_0(t,0)$	$\hat{U}_I(t,0)\hat{A}_H(t)\hat{U}_I^\dagger(t,0)$	

analogous to (2.24). In perturbation theory, the series is truncated at finite n and in case of $n = 1$, we get

$$|\Psi_I^1(t)\rangle = |\Psi(0)\rangle - \frac{i}{\hbar}\int_0^t dt'\hat{U}_0^\dagger(t',0)\hat{W}(t')\hat{U}_0(t',0)|\Psi(0)\rangle. \quad (2.82)$$

Going back to the Schrödinger picture, by using (2.73), we get

$$|\Psi_S^1(t)\rangle = \hat{U}_0(t,0)|\Psi(0)\rangle - \frac{i}{\hbar}\int_0^t dt'\hat{U}_0(t,t')\hat{W}(t')\hat{U}_0(t',0)|\Psi(0)\rangle. \quad (2.83)$$

in first order. We will use expressions of this kind in the discussion of pump–probe spectroscopy in Chap. 5. Terms of higher order will contain multiple integrals but will not be needed there.

2.2.4 Magnus Expansion

Another approach to solve the time-dependent Schrödinger equation starting from the time-evolution operator is the so-called Magnus expansion. The basic idea of this method is resummation and can be understood by considering a function depending on a parameter λ. Expanding this function in powers of the parameter leads to

$$A = A_0(1 + \lambda A_1 + \lambda^2 A_2 + \ldots). \quad (2.84)$$

Alternatively, the function can be written as a prefactor times an exponential

$$A = A_0\exp(F). \quad (2.85)$$

Also the exponent F can be expanded in powers of the parameter

$$F = \lambda F_1 + \lambda^2 F_2 + \dots \tag{2.86}$$

The exponential function can now be Taylor expanded and after comparing the coefficients of equal powers of λ, the F_n can be expressed in terms of the A_n. If we now truncate the series in (2.84) after $n = 2$, by using the coefficients in the exponent via

$$A = A_0 \exp(\lambda A_1 + \lambda^2(A_2 - A_1^2/2)), \tag{2.87}$$

an expression of infinite order in λ has been gained.

In quantum dynamics, the technique presented above is used for the infinite sum in (2.81), representing the time-evolution operator. The parameter $\lambda = -i/\hbar$ in this case. The final result for the time-evolution operator is

$$\hat{U}_I(t,0) = \hat{T} \exp\left\{ -\frac{i}{\hbar} \int_0^t dt' \hat{W}_I(t') \right\}$$
$$= \exp\left\{ \sum_{n=1}^{\infty} \frac{1}{n!} \left(-\frac{i}{\hbar} \right)^n \hat{H}_n(t,0) \right\}, \tag{2.88}$$

where

$$\hat{H}_1 = \int_0^t dt' \hat{W}_I(t') \tag{2.89}$$

$$\hat{H}_2 = \int_0^t dt_2 \int_0^{t_2} dt_1 [\hat{W}_I(t_2), \hat{W}_I(t_1)] \tag{2.90}$$

$$\hat{H}_3 = \int_0^t dt_3 \int_0^{t_3} dt_2 \int_0^{t_2} dt_1 ([\hat{W}_I(t_3), [\hat{W}_I(t_2), \hat{W}_I(t_1)]]$$
$$+ [[\hat{W}_I(t_3), \hat{W}_I(t_2)], \hat{W}_I(t_1)]) \tag{2.91}$$

are the first three terms in the expansion of the exponent.

Exercise 2.8 *Prove the second order expression $\hat{H}_2$ of the Magnus expansion in the exponent of the time-evolution operator in the interaction picture.*

The main advantage of the expression in (2.88) is that in principle it is an exact result and that it does not contain the time-ordering operator any more. In numerical applications the summation in the exponent will be terminated at finite order n, however, and leads to a unitary propagation scheme at any order. If one would truncate the expansion after $n = 1$, then the time-ordering operator in (2.88) would have been ignored altogether. Although this seems to be rather a crude approximation, in Chap. 4 we will see that it leads to reasonable results in the case of atoms subject to extremely short pulses. Furthermore, it has turned out that in the interaction picture with a suitable choice of $\hat{H}_0$, truncating the Magnus expansion is a successful numerical approach [17].

2.2.5 Time-Dependent Hartree Method

Especially in Chap. 5, we will investigate systems with several coupled degrees of freedom. The factorization of the total wavefunction is the first very crude approximative way to solve the time-dependent Schrödinger equation for such systems. It shall therefore be discussed here for the simplest case of two (distinguishable) degrees of freedom.

The total Hamiltonian shall be of the form

$$\hat{H} = \sum_{n=1}^{2} \hat{H}_n(x_n) + V_{12}(x_1, x_2) \tag{2.92}$$

with single particle operators

$$\hat{H}_n(x_n) = -\frac{\hbar^2}{2m} \frac{\partial^2}{\partial x_n^2} + V_n(x_n) \tag{2.93}$$

and the coupling potential V_{12} depending on the two coordinates in a non-additive manner. The so-called Hartree Ansatz for the wavefunction is of the form

$$\Psi(x_1, x_2, t) = \Psi_1(x_1, t)\Psi_2(x_2, t) \tag{2.94}$$

of a product of single particle wavefunctions.

This Ansatz is exact in the case that the coupling V_{12} vanishes. The single particle functions then fulfill

$$i\hbar\dot{\Psi}_n(x_n, t) = \hat{H}_n\Psi_n(x_n, t). \tag{2.95}$$

We now plug the Hartree Ansatz into the full time-dependent Schrödinger equation and find

$$i\hbar\left(\Psi_2\dot{\Psi}_1 + \Psi_1\dot{\Psi}_2\right) = \Psi_2\hat{H}_1\Psi_1 + \Psi_1\hat{H}_2\Psi_2 + V_{12}\Psi_1\Psi_2. \tag{2.96}$$

Multiplying this equation with Ψ_2^* and integrating over the coordinate of particle 2 yields

$$i\hbar\left(\dot{\Psi}_1 + \Psi_1\langle\Psi_2|\dot{\Psi}_2\rangle_2\right) = \hat{H}_1\Psi_1 + \Psi_1\langle\Psi_2|\hat{H}_2|\Psi_2\rangle_2 + \langle\Psi_2|V_{12}|\Psi_2\rangle_2\Psi_1. \tag{2.97}$$

By using the single particle equations of zeroth order with the index 2, the second terms on the LHS and the RHS cancel each other and one finds

$$i\hbar\dot{\Psi}_1(x_1, t) = \left(-\frac{\hbar^2}{2m}\Delta_1 + V_{1,\text{eff}}(x_1, t)\right)\Psi_1(x_1, t) \tag{2.98}$$

with an effective, time-dependent potential

$$V_{1,\text{eff}}(x_1, t) = V_1(x_1) + \langle\Psi_2|V_{12}|\Psi_2\rangle_2. \tag{2.99}$$

An analogous equation can be derived for particle 2

$$i\hbar\dot{\Psi}_2(x_2,t) = \left(-\frac{\hbar^2}{2m}\Delta_2 + V_{2,\text{eff}}(x_2,t)\right)\Psi_2(x_2,t), \qquad (2.100)$$

by multiplying the time-dependent Schrödinger equation with Ψ_1^* and integrating over x_1.

The particles move in effective "mean" fields that are determined by the dynamics of the other particle. The coupled equations have to be solved self-consistently. This is the reason that the Hartree method sometimes is called a TDSCF (time-dependent self consistent field) method. The multi-configuration time-dependent Hartree (MCTDH) method [18] goes far beyond what has been presented here and in principle allows for an exact numerical solution of the time-dependent Schrödinger equation.

2.2.6 Quantum-Classical Methods

In quantum-classical methods the degrees of freedom are separated into a subset that shall be dealt with on the basis of classical mechanics and a subset to be described fully quantum mechanically. Analogously to the Hartree method, the classical degrees of freedom will move in an effective potential that is determined by the solution of the quantum problem.

For reasons of convenience, we start the discussion with the time-independent Schrödinger equation and restrict it to the case of two degrees of freedom: a light particle with coordinate x and mass m and a heavy one with X and M, respectively. Having the full solution of the time-independent Schrödinger equation

$$\hat{H}\psi_n(x,X) = E_n\psi_n(x,X) \qquad (2.101)$$

with the Hamiltonian

$$\hat{H} = \frac{\hat{p}^2}{2m} + \frac{\hat{P}^2}{2M} + v(x,X) + V(X) \qquad (2.102)$$

at hand would allow us to construct a time-dependent solution according to

$$\Psi(x,X,t) = \sum_n c_n \exp\left[-\frac{i}{\hbar}E_n t\right]\psi_n(x,X). \qquad (2.103)$$

A way to treat the coupled system approximately is intimately related to the separation Ansatz of Sect. 2.2.5 and will be discussed in much more detail later-on in the section on the Born–Oppenheimer approximation. The idea is simple: one replaces the coupled problem by a pair of uncoupled single particle problems. In order to do so, first the light particle is dealt with under certain (fixed) conditions for the heavy particle (denoted by $|X$)

$$\hat{H}^0(x|X)\phi_j(x|X) = \epsilon_j^0(X)\phi_j(x|X), \qquad (2.104)$$

where

$$\hat{H}^0(x|X) = \frac{\hat{p}^2}{2m} + v(x, X) \qquad (2.105)$$

depends parametrically on X and j is the quantum number of the light particle.

Using the product Ansatz

$$\psi_n(x, X) \approx \phi_j(x|X)\chi_{l,j}(X) \qquad (2.106)$$

in the full time-independent Schrödinger equation and integrating out the coordinate of the light particle in the same way as in Sect. 2.2.5, one arrives at equations of the form

$$\hat{H}_j^1(X)\chi_{l,j}(X) = \epsilon_{l,j}^1\chi_{l,j}(X) \qquad (2.107)$$

with the Hamiltonian

$$\hat{H}_j^1(X) = \frac{\hat{P}^2}{2M} + V(X) + \epsilon_j^0(X) \qquad (2.108)$$

and eigenvalues which are approximately given by

$$E_n \approx \epsilon_{l,j}^1 \qquad (2.109)$$

depending on j as well as on l, which is the quantum number of the heavy particle. The heavy particle is thus governed by an effective potential, depending on the quantum state of the light particle.

Let us now turn to dynamics. In the Ehrenfest method one postulates the classical treatment of the heavy particle. Analogous to the effective potential an effective force can be derived which reads

$$F_{\text{eff}} = -\frac{\partial}{\partial X}\left\{V + \int dx \Phi^* \hat{H}^0 \Phi\right\}, \qquad (2.110)$$

and where the wavefunction of the light particle fulfills the time-dependent Schrödinger equation

$$i\hbar\dot{\Phi}(x|X(t), t) = \hat{H}^0(x|X(t))\Phi(x|X(t), t). \qquad (2.111)$$

Expanding this wavefunction in eigenfunctions of the light particle

$$\Phi(x|X(t), t) = \sum_j c_j(t)\phi_j(x|X(t)) \qquad (2.112)$$

yields coupled differential equations for the coefficients

$$i\hbar \dot{c}_j(t) = \epsilon_j^0 c_j - i\hbar \dot{X} \sum_k d_{j,k} c_k, \qquad (2.113)$$

where $d_{jk} = \int dx \phi_j \frac{\partial \phi_k}{\partial X}$. Together with the explicit expression

$$F_{\text{eff}} = -\frac{\partial V}{\partial X} - \sum_j |c_j|^2 \frac{\partial \epsilon_j^0}{\partial X} + \sum_{j,k<j} [c_j^* c_k + c_k^* c_j][\epsilon_j^0 - \epsilon_k^0] d_{j,k} \qquad (2.114)$$

for the effective force, this can be proven by working through Exercise 2.9.

Exercise 2.9 *Verify the fundamental equations of the Ehrenfest method.*

(a) First prove the validity of the coupled differential equations for the coefficients c_j (Use the product and the chain rule of differentiation).
(b) Calculate the effective force by using $d_{kj} = -d_{jk}$ (Proof?)

The first term in the expression of the force is the so-called external force, whereas the second one describes adiabatic and the third one nonadiabatic dynamics.[7] The last two terms have to be determined by solving the quantum problem of the light particle. An alternative quantum-classical approach is the so-called surface hopping technique. Its relation to the Ehrenfest approach, and which method is suited under which circumstances are discussed in [19].

2.2.7 Floquet Theory

In dealing with laser driven systems the problem of time-periodic Hamiltonians is of central importance. In this case, we have

$$\hat{H}(t+T) = \hat{H}(t) \qquad (2.115)$$

with the period $T = 2\pi/\omega$ of the external perturbation.

As in the general time-dependent case, the time evolution operator can be used to propagate a wavefunction. In addition to the properties discussed in Sect. 2.1.2 we can now make use of the property

$$\hat{U}(t+T, s+T) = \hat{U}(t, s). \qquad (2.116)$$

To solve the time-dependent Schrödinger equation we prove that the Hamiltonian extended by the time derivative

$$\hat{\mathcal{H}}(t) \equiv \hat{H}(t) - i\hbar \partial_t \qquad (2.117)$$

commutes with the time-evolution operator over one period. The time-dependent Schrödinger equation can be rewritten by using the above definition and we apply the time-evolution operator from the left

---

[7] The explanation of these terms follows in Chap. 5.

$$\hat{U}(t+T,t)\hat{\mathcal{H}}(t)|\Psi(t)\rangle = 0 \qquad (2.118)$$

$$\hat{U}(t+T,t)\hat{\mathcal{H}}(t)\hat{U}^{-1}(t+T,t)\hat{U}(t+T,t)|\Psi(t)\rangle = 0 \qquad (2.119)$$

$$\hat{\mathcal{H}}(t)\hat{U}(t+T,t)|\Psi(t)\rangle = 0. \qquad (2.120)$$

The last, decisive step follows form the periodicity of the Hamiltonian and with the help of the chain rule.[8] We can thus conclude that

$$[\hat{U}(t+T,t),\hat{\mathcal{H}}(t)] = 0 \qquad (2.121)$$

holds. The two operators have a common system of eigenfunctions which shall be denoted by $|\Psi_\epsilon(t)\rangle$.

From the composition property (2.20) and from (2.116), we can conclude that

$$\hat{U}(2T,0) = \hat{U}(2T,T)\hat{U}(T,0) = \hat{U}^2(T,0). \qquad (2.122)$$

The group of time-evolution operators over one period therefore is an Abelian group. Its eigenfunctions have to transform according to a one-dimensional irreducible representation [20]

$$\hat{U}(T,0)|\Psi_\epsilon(0)\rangle = \exp\left\{-\frac{i}{\hbar}\epsilon T\right\}|\Psi_\epsilon(0)\rangle. \qquad (2.123)$$

Comparing this equation with the time-evolution over one period

$$\hat{U}(T,0)|\Psi_\epsilon(0)\rangle = |\Psi_\epsilon(T)\rangle \qquad (2.124)$$

leads to the Floquet theorem for the solution of the time-dependent Schrödinger equation

$$\Psi_\epsilon(x,t) = \exp\left\{-\frac{i}{\hbar}\epsilon t\right\}\psi_\epsilon(x,t) \qquad (2.125)$$

$$\psi_\epsilon(x,t) = \psi_\epsilon(x,t+T). \qquad (2.126)$$

The wavefunction is a product of an exponential factor times a periodic function.[9] The factor ϵ in the exponent of (2.125) is sometimes referred to as *Floquet exponent* and the corresponding periodic function ψ_ϵ is called *Floquet function* to honor the French mathematician Gaston Floquet, who worked on differential equations with periodic coefficients in the nineteenth century.

The product in (2.125) is formally analogous to the separation (2.7) in the stationary case. To use this analogy, we rewrite the time-dependent Schrödinger equation as in (2.118)

$$\hat{\mathcal{H}}(x,t)\Psi(x,t) = 0. \qquad (2.127)$$

---

[8] We have used $\hat{U}(t+T,t)\hat{\mathcal{H}}(t)\hat{U}^{-1}(t+T,t) = \hat{\mathcal{H}}(t+T) = \hat{H}(t+T) - i\hbar\partial_{t+T} = \hat{H}(t) - i\hbar\partial_t$.

[9] Formally, this Ansatz is equivalent to the Bloch theorem of solid-state physics.

Inserting the Floquet solution (2.125) and performing the time-derivative of the exponential part yields

$$\hat{\mathcal{H}}(x,t)\psi_\alpha(x,t) = \epsilon_\alpha \psi_\alpha(x,t). \tag{2.128}$$

This "Floquet type Schrödinger equation" has the same formal structure as the time-independent Schrödinger equation. Therefore, the Floquet exponents are also called *quasi-energies* and the Floquet functions are referred to as *quasi-eigenfunctions*. In the case of a Hermitian Hamiltonian the quasi-energies are real.

Exercise 2.10 *Use the extended scalar product*

$$\langle\langle u_\alpha | v_\beta \rangle\rangle = \frac{1}{T}\int_0^T dt \int_{-\infty}^{\infty} dx u_\alpha^*(x,t) v_\beta(x,t)$$

and the extended Hamiltonian $\hat{\mathcal{H}} = \hat{H}(t) - i\hbar\partial_t$ with $\hat{H}(t+T) = \hat{H}(t)$, to show that the Floquet energies ϵ_α are real in case of Hermitian Hamiltonias $\hat{H}(t)$.

In the case of vanishing external field the Hamiltonian becomes time-independent. This implies that also ψ_α is time-independent. The index α, therefore, is related to the quantum numbers of the unperturbed problem. It is a special feature of the Floquet solution (2.125), (2.126) that the modified quasi-eigenfunctions and corresponding energies

$$\psi_{\alpha'}(x,t) := \psi_\alpha(x,t)\exp(ik\omega t) \tag{2.129}$$
$$\epsilon_{\alpha'} := \epsilon_\alpha + k\hbar\omega \tag{2.130}$$

with $k = 0, \pm 1, \pm 2, \ldots$[10] are equivalent to $\psi_\alpha(x,t), \epsilon_\alpha$, due to the fact that they correspond to the same total solution $\Psi_\alpha(x,t)$. The index

$$\alpha' := (\alpha, k) \quad k = 0, \pm 1, \pm 2, \ldots \tag{2.131}$$

denotes a class of infinitely many solutions. Out of each class only one lays in a so-called Brillouin zone of width $\hbar\omega$, however. The discussion given above and more details on the underlying Hilbert space theory can be found in [21]. Without proving the completeness, we will use

$$\hat{1} = \sum_{\alpha'} |\psi_{\alpha'}(0)\rangle\langle\psi_{\alpha'}(0)| \tag{2.132}$$

as the representation of unity in terms of (discrete) Floquet states. For this representation to be true also in the nondriven case, it is clear that only one member of the class of equivalent eigensolutions may contribute to the sum given above. A solution of the time-dependent Schrödinger equation can

---

[10] Note that k has to be an integer in order for the modified quasi-eigenfunction to be periodic.

therefore be written as a superposition of quasi-eigenfunctions with appropriate coefficients

$$|\Psi(t)\rangle = \sum_{\alpha'} c_{\alpha'} |\psi_{\alpha'}(t)\rangle \exp\left\{-\frac{i}{\hbar}\epsilon_{\alpha'}t\right\} \tag{2.133}$$

$$c_{\alpha'} = \langle\psi_{\alpha'}(0)|\Psi(0)\rangle. \tag{2.134}$$

This equation exhibits clearly that the quasi-energies determine the long-term time-evolution of a periodically driven quantum system. The behavior of the quasi-energies as a function of an external parameter (e.g., the field strength or the frequency) will therefore be very important. To study this behavior, the symmetry of the Hamiltonian will be decisive. We will come back to this point in Appendix 3.A to Chap. 3.

2.3 Numerical Methods

Apart from special two-level problems that will be dealt with in Chap. 3 and systems with maximally quadratic potentials (and problems that can be mapped onto such cases) there are only a few exactly analytically solvable problems in quantum dynamics, as can be seen by studying the review by Kleber [22].

Almost all interesting problems of atomic and molecular physics with and without the presence of laser fields display classically nonlinear dynamics, however. Exact numerical solutions of the quantum dynamics are therefore sought for. Apart from time-dependent information that is needed for the description of e.g. pump–probe experiments, to be discussed in Chap. 5, also spectral information for systems with autonomous Hamiltonians can be gained from time series as was shown in Sect. 2.1.2.

In the following, different ways to solve the time-dependent Schrödinger equation numerically will be described. First, we will review some numerically exact methods and in the end the implementation of the semiclassical theory lined out in Sect. 2.2.2. All methods can be characterized according to two formal criteria, which will be called problem (a), respectively, problem (b) in the following:

(a) Which (finite) basis is used to represent the wavefunction?
(b) In which (approximate) way is the time-evolution performed?

We will distinguish the methods according to their different approach to the solution of the problems above.

2.3.1 Orthogonal Basis Expansion

All methods to solve the time-dependent Schrödinger equation numerically have to deal with a way to express the wavefunction in a certain representation. Here we shall consider the expansion of the wavefunction in a set of orthogonal basis functions, which are eigenfunctions of a certain (simple) Hamiltonian, as e.g., the harmonic oscillator one

$$\hat{H}_{\text{HO}} = -\frac{\hbar^2}{2m}\frac{\partial^2}{\partial x^2} + \frac{1}{2}m\omega_e^2 x^2. \tag{2.135}$$

Its eigenfunctions are given by

$$\psi_n(x) = \sqrt{\frac{\sigma}{n!2^n\sqrt{\pi}}}H_n(\sigma x)\exp\left\{-\frac{1}{2}\sigma^2 x^2\right\} \tag{2.136}$$

where H_n are Hermite polynomials [23] and $\sigma = \sqrt{m\omega_e/\hbar}$.

The alternative representation of the harmonic oscillator Hamiltonian

$$\hat{H}_{\text{HO}} = \hbar\omega_e\left(\hat{a}^\dagger\hat{a} + \frac{1}{2}\right) \tag{2.137}$$

in terms of so-called creation and annihilation operators

$$\hat{a}^\dagger = \frac{1}{\sqrt{2}}\left(\sigma\hat{x} - \frac{1}{\sigma}\frac{\partial}{\partial x}\right) \tag{2.138}$$

$$\hat{a} = \frac{1}{\sqrt{2}}\left(\sigma\hat{x} + \frac{1}{\sigma}\frac{\partial}{\partial x}\right) \tag{2.139}$$

is very helpful. These operators fulfill the commutation relation

$$[\hat{a}, \hat{a}^\dagger] = \hat{1} \tag{2.140}$$

and have the properties

$$\hat{a}^\dagger|n\rangle = \sqrt{n+1}|n+1\rangle \qquad n = 0, 1, 2, \ldots \tag{2.141}$$

$$\hat{a}|n\rangle = \sqrt{n}|n-1\rangle \qquad n = 1, 2, \ldots \tag{2.142}$$

i.e., they either allow to "climb up" or "climb down" the ladder of harmonic oscillator states (in short-hand notation just denoted by the index n) and are therefore sometimes also referred to as ladder operators.

By resolving the definition of the ladder operators in terms of the position and the derivative operator

$$\hat{x} = \frac{1}{\sqrt{2}\sigma}(\hat{a} + \hat{a}^\dagger) \tag{2.143}$$

$$\frac{\partial}{\partial x} = \frac{\sigma}{\sqrt{2}}(\hat{a} - \hat{a}^\dagger), \tag{2.144}$$

one can express arbitrary powers of these operators in terms of products of $\hat{a}^\dagger$ and $\hat{a}$. Matrix elements of any operator between harmonic oscillator states can then be calculated by employing (2.141) and (2.142).

An arbitrary time-dependent wavefunction can now be expanded into eigenfunctions of the harmonic oscillator according to

$$|\Psi(t)\rangle = \sum_{l=0}^{\infty} d_l(t)|l\rangle. \tag{2.145}$$

After insertion of this expression into the time-dependent Schrödinger equation governed by the Hamiltonian $\hat{H}$, an infinite linear system of coupled ordinary differential equations for the expansion coefficients

$$i\hbar\dot{d}_n(t) = \sum_{l=0}^{\infty} d_l(t)\langle n|\hat{H}|l\rangle \tag{2.146}$$

is gained. This system could (in principle) be solved if the initial conditions were known. At this point, however, we should address our "problems" as stated in the introduction to this section:

(a) The basis problem is solved by truncating the expansion at a large $l = L-1$, which is e.g., determined by the initial state that shall be described. One thus uses a "Finite Basis Representation". Convergence of the results can be checked by increasing the size L of the finite basis.
(b) The numerical integration of the linear system of differential equations could be performed with the help of a suitable integration routine, as e.g., the Runge–Kutta method [24].

Solving the system of coupled differential equations can be circumvented, however, by finding the (first N) eigenvalues E_n and eigenfunctions $|n_H\rangle$ of the Hamiltonian $\hat{H}$ in case this is autonomous. By determining the (now time-independent) expansion coefficients of the wavefunction in terms of these eigenfunctions, the wavefunction is easily time-evolved by using the corresponding eigenenergies according to

$$|\Psi(t)\rangle = \sum_{n=0}^{N} c_n|n_H\rangle e^{-iE_n t/\hbar}. \tag{2.147}$$

We should keep in mind, however, that the solution of the eigenvalue problem requires a numerical effort of order L^3 if $L \times L$ is the size of the matrix to be diagonalized [24] and is therefore only suitable if L can be kept small.

The Floquet Matrix

The second method to solve problem (b) is fortunately not restricted to autonomous Hamiltonians. It also works for periodically driven systems that have been discussed in Sect. 2.2.7 and leads to the calculation of the quasi-energies and quasi-eigenfunctions. We start from the Floquet type Schrödinger equation in bracket notation

$$\hat{\mathcal{H}}(t)|\psi_\alpha(t)\rangle = \epsilon_\alpha|\psi_\alpha(t)\rangle \tag{2.148}$$

with the extended Hamiltonian defined in (2.117). Due to the periodic time-dependence of the Floquet functions they can be Fourier expanded according to

$$|\psi_\alpha(t)\rangle = \sum_{n=-\infty}^{\infty} |\psi_\alpha^n\rangle e^{-in\omega t}. \tag{2.149}$$

The Fourier coefficients on the RHS of this expression can in turn be expanded in an orthogonal system $\{|k\rangle\}$

$$|\psi_\alpha^n\rangle = \sum_{k=0}^{\infty} \psi_{k,\alpha}^n |k\rangle \tag{2.150}$$

and the Schrödinger equation thus is given by

$$\sum_{n=-\infty}^{\infty}\sum_{k=0}^{\infty} \hat{\mathcal{H}}\psi_{k,\alpha}^n |k\rangle e^{-in\omega t} = \sum_{n=-\infty}^{\infty}\sum_{k=0}^{\infty} \epsilon_\alpha \psi_{k,\alpha}^n |k\rangle e^{-in\omega t}. \tag{2.151}$$

Multiplying this equation with $\langle lm| := \langle \psi_l e^{-im\omega t}|$ from the left and integrating over one period of the external force yields

$$\sum_{n=-\infty}^{\infty}\sum_{k=0}^{\infty} \langle\langle lm|\hat{\mathcal{H}}|kn\rangle\rangle \psi_{k,\alpha}^n = \epsilon_\alpha \psi_{l,\alpha}^m, \tag{2.152}$$

where we have used the definition $\langle\langle\ldots\rangle\rangle = \frac{1}{T}\int_0^T dt\langle\ldots\rangle$ of the extended scalar product, that has already been used in Exercise 2.10.

Equation (2.152) has first been given by Shirley [25] in the case of a two-level system. Shirley then transformed the equations to recover an eigenvalue problem, whose solutions are the quasi-energies. This procedure can also be used for an infinite dimensional Hilbert space, however. To do so, one rewrites the equation above according to

$$\sum_{n=-\infty}^{\infty}\sum_{k=0}^{\infty} \left\{ \langle l|\hat{H}^{[m-n]}|k\rangle - n\hbar\omega\delta_{mn}\delta_{lk} \right\} \psi_{k,\alpha}^n = \epsilon_\alpha \psi_{l,\alpha}^m \tag{2.153}$$

where the definition

$$\hat{H}^{[m-n]} = \frac{1}{T}\int_0^T dt \hat{H}(t)\exp(i[m-n]\omega t) \tag{2.154}$$

has been introduced. In the case of a monochromatic perturbation

$$\hat{H}(t) \equiv \hat{H}_0 + \hat{H}_1\sin(\omega t) \tag{2.155}$$

time integration yields

$$\hat{H}^{[m-n]} = \hat{H}_0 \delta_{m,n} + \frac{\hat{H}_1}{2\mathrm{i}} \left\{ \delta_{mn-1} - \delta_{mn+1} \right\}. \tag{2.156}$$

Equation (2.152) is the eigenvalue problem of the extended Hamiltonian $\hat{\mathcal{H}}$, whose matrix elements are given by

$$\langle l | \hat{H}^{[m-n]} | k \rangle - n\hbar\omega \delta_{mn} \delta_{lk}. \tag{2.157}$$

The Fourier expansion (2.149) has rendered the problem time-independent. One has to cope with an additional "dimension" ($n = 0, \pm 1, \pm 2 \ldots$), however.

After choosing a basis (e.g., the harmonic basis) and calculating the matrix elements, the quasi-energies are the eigenvalues and the quasi-eigenfunctions are the eigenvectors of (2.153). The Floquet matrix to be diagonalized is

$$\begin{pmatrix}
\ddots & & & & & \\
& \mathbf{H_0} - 2\hbar\omega\mathbf{1} & \frac{1}{2\mathrm{i}}\mathbf{H_1} & 0 & 0 & 0 \\
& -\frac{1}{2\mathrm{i}}\mathbf{H_1} & \mathbf{H_0} - 1\hbar\omega\mathbf{1} & \frac{1}{2\mathrm{i}}\mathbf{H_1} & 0 & \\
& 0 & -\frac{1}{2\mathrm{i}}\mathbf{H_1} & \mathbf{H_0} & \frac{1}{2\mathrm{i}}\mathbf{H_1} & 0 \\
& 0 & 0 & -\frac{1}{2\mathrm{i}}\mathbf{H_1} & \mathbf{H_0} + 1\hbar\omega\mathbf{1} & \frac{1}{2\mathrm{i}}\mathbf{H_1} \\
& 0 & 0 & 0 & -\frac{1}{2\mathrm{i}}\mathbf{H_1} & \mathbf{H_0} + 2\hbar\omega\mathbf{1} \\
& & & & & \ddots
\end{pmatrix}. \tag{2.158}$$

Here $\mathbf{1}, \mathbf{0}$ are unit and zero matrices, and in principle the block matrices have to be added ad infinitum, i.e., $n \to \infty$. In the numerics, however, one uses matrices $\mathbf{H_0}$ and $\mathbf{H_1}$ of finite size $L \times L$ as well as a finite number $2M + 1$ of Fourier terms. Convergence can be checked by increasing L as well as M.

In general, the basis function expansion method is only a viable approach if the matrix elements of the Hamiltonian can be calculated easily. If the basis is the harmonic oscillator one, this is the case if the potential e.g., is given by a polynomial of low order. In other cases or if the potential is multidimensional, so-called "discrete variable representations" (DVR) [26] are frequently used. Finally, it should be noted that the diagonalization of the Floquet matrix becomes much more difficult, if the system under consideration contains a continuum of states. Then the method of complex rotation can be employed [27].

2.3.2 Split-Operator FFT Method

The split-operator method for the solution of the time-dependent Schrödinger equation is based on the approximate representation of the time-evolution operator, i.e., the treatment of problem (b) by using the Zassenhaus formula [28][11]

---

[11] The Baker–Campbell–Haussdorff (BCH) formula is the dual relation and reads
$$e^{\hat{x}} e^{\hat{y}} = e^{\hat{x}+\hat{y}+1/2[\hat{x},\hat{y}]+1/12([\hat{x},[\hat{x},\hat{y}]]+[\hat{y},[\hat{y},\hat{x}]])+\cdots}.$$

$$e^{\hat{x}+\hat{y}} = e^{\hat{x}}e^{\hat{y}}e^{-1/2[\hat{x},\hat{y}]}e^{1/3[\hat{y},[\hat{x},\hat{y}]]+1/6[\hat{x},[\hat{x},\hat{y}]]}\ \dots \tag{2.159}$$

If the Hamiltonian is of the usual form $\hat{H} = \hat{T}_k + \hat{V}$, for very short time steps Δt, one finds from the Zassenhaus formula that

$$e^{-i\hat{H}\Delta t/\hbar} = e^{-i\hat{T}_k\Delta t/\hbar}e^{-i\hat{V}\Delta t/\hbar} + O(\Delta t^2) \tag{2.160}$$

holds. This approximation is also referred to as the Trotter product. By working through Exercise 2.11 you can convince yourself that a more "symmetric" splitting of the Hamiltonian according to

$$e^{-i\hat{H}\Delta t/\hbar} = e^{-i\hat{V}\Delta t/(2\hbar)}e^{-i\hat{T}_k\Delta t/\hbar}e^{-i\hat{V}\Delta t/(2\hbar)} + O(\Delta t^3) \tag{2.161}$$

leads to a formula of higher accuracy.

Exercise 2.11 *Show that the symmetric splitting of the time-evolution operator is correct up to order $O(\Delta t^3)$.*
Hint: Use the Zassenhaus as well as the BCH formula

Problem (a) is now dealt with by representing the wavefunction at $t = 0$ on a position space grid $x_n \in [x_{\min}, x_{\max}], n = 1, \dots, N$. The wavefunction propagated for a time Δt at the grid point x_n is then given by

$$\Psi(x_n, \Delta t) = \langle x_n|e^{-i\hat{H}\Delta t/\hbar}|\Psi(0)\rangle$$
$$= \langle x_n|e^{-i\hat{V}\Delta t/(2\hbar)}e^{-i\hat{T}_k\Delta t/\hbar}e^{-i\hat{V}\Delta t/(2\hbar)}|\Psi(0)\rangle. \tag{2.162}$$

By inserting unity twice in terms of position states and once in terms of momentum states, the threefold integral (for the numerics, the integrations are discretized due to the grid based representation of the wavefunction)

$$\Psi(x_n, \Delta t) = \int \mathrm{d}x' \int \mathrm{d}p' \int \mathrm{d}x'' \langle x_n|e^{-i\hat{V}\Delta t/(2\hbar)}|x''\rangle$$
$$\langle x''|e^{-i\hat{T}_k\Delta t/\hbar}|p'\rangle\langle p'|e^{-i\hat{V}\Delta t/(2\hbar)}|x'\rangle\langle x'|\Psi(0)\rangle \tag{2.163}$$

emerges. The integral over x'' can be performed immediately due to the locality of the potential in position space and the δ function appearing in

$$\langle x_n|e^{-i\hat{V}\Delta t/(2\hbar)}|x''\rangle = e^{-iV(x_n)\Delta t/(2\hbar)}\delta(x'' - x_n). \tag{2.164}$$

Also the second exponentiated potential term simplifies due to locality according to

$$\langle p'|e^{-i\hat{V}\Delta t/(2\hbar)}|x'\rangle = \langle p'|x'\rangle e^{-iV(x')\Delta t/(2\hbar)}$$
$$= \frac{1}{\sqrt{2\pi\hbar}}e^{-ip'x'/\hbar}e^{-iV(x')\Delta t/(2\hbar)}. \tag{2.165}$$

The x' integration is a Fourier transformation of the "intermediate wavefunction" into momentum space. This leads to the fact that the exponentiated operator of kinetic energy becomes local and can be applied easily via

$$\langle x''|e^{-i\hat{T}_k\Delta t/\hbar}|p'\rangle = \langle x''|p'\rangle e^{-iT_k(p')\Delta t/\hbar}$$

$$= \frac{1}{\sqrt{2\pi\hbar}}e^{ip'x''/\hbar}e^{-iT_k(p')\Delta t/\hbar}. \qquad (2.166)$$

The p' integration transforms the wavefunction back into position space.

The main numerical effort is the need to perform two Fourier transforms of the wavefunction during the propagation over one time step. These can be performed by using the Fast Fourier Transformation (FFT) algorithm [24], however. The implementation of the split-operator based FFT method[12] can therefore be summarized as follows:

   (i) Represent the initial wavefunction on a position space grid
  (ii) Apply the local operator $e^{-i\hat{V}\Delta t/(2\hbar)}$
 (iii) Do an FFT into momentum space
 (iv) Apply the local operator $e^{-i\hat{T}_k\Delta t/\hbar}$
  (v) Do an inverse FFT back into position space
 (vi) Apply the local operator $e^{-i\hat{V}\Delta t/(2\hbar)}$.

This procedure is applied for the propagation over a very small time step. For the propagation over a finite time it will be repeated frequently and if the intermediate values of the wavefunction are not needed the two half time steps of potential propagation can be combined (apart from the first and the last one). Furthermore, to propagate the wavefunction over the next time step, we will need its value not only at x_n but at all values of x. This is reflecting the *nonlocal* nature of quantum theory. For the calculation of the new wavefunction the old one is needed everywhere. This is in contrast to classical mechanics. A trajectory only depends on its initial conditions; classical mechanics is a local theory.

A nice review of the details of FFT and a corresponding subroutine can be found in [24]. Some facts will be briefly repeated here. The discrete version of the Fourier transform is

$$\Phi(x_i) = \sum_{k=-N/2-1}^{N/2} a_k e^{2\pi i k x_i/X} \qquad (2.167)$$

$$a_k = \frac{1}{N}\sum_{n=1}^{N} \Phi(x_n)e^{-2\pi i k x_n/X}. \qquad (2.168)$$

For implementation it is important to note that

- $N = 2^j$ has to be a power of 2
- The grid length is $X = x_{\max} - x_{\min}$ and x_n are equidistant with $\Delta x = X/N$
- The numerical effort scales with $N \ln N$ [24]

---

[12] Originally this approach was proposed by Fleck, Morris and Feit for the solution of the Maxwellian wave-equation [29].

- The maximal momentum that can be described is

$$p_{\max} = h/(2\Delta x) = Nh/(2X),$$

and $p_{\min} = -p_{\max}$
- The covered phase space volume is $V_P = 2X p_{\max} = Nh$
- The time step should fulfill $\Delta t < \pi/(3E_{\max})$, with $E_{\max}$ the maximal energy to be described [30]. For very long propagation times see also [31].
- Energy resolution is given by $\Delta E_{\min} = \pi/T_t$, with T_t the total propagation time.

There are more recent implementations of FFT which do not have the restriction to integer powers of 2 and which, through adaption to the platform that is used for the calculations can have considerable advantages in speed (see e.g., FFTW: Fastest Fourier Transform in the West [32]).

Finally, it should be mentioned that the usage of the nontime-ordered time-evolution operator at the beginning of our discussion is no restriction to time-independent Hamiltonians. As in the case of the infinitesimal time-evolution operator, one can use a constant Hamiltonian for the propagation over a small time interval. At the beginning of the next time step a slightly changed Hamiltonian is employed. One drawback of the method, that will not come into play in the present book, however, shall be mentioned. The split-operator idea only succeeds in producing local operators if there are no products of $\hat{p}$ and $\hat{x}$ in the Hamiltonian. These would appear in the treatment of dissipative quantum problems, which are outside the scope of this presentation.

Negative Imaginary Absorbing Potentials

Another possible drawback of a grid based method like the split-operator FFT method shall be dealt with in a bit more detail. It can be cast in the form of a question: What happens to a wavepacket, when it hits the grid boundaries? It would reenter on the other side of the grid, leading to nonphysical results! This can be avoided by adding a negative imaginary potential of the form

$$V(x) = -if(x)\Theta(x - x_a), \tag{2.169}$$

which is nonzero for values $x > x_a$, close to the right grid boundary $x_{\max}$ and a similar term at the left side of the grid. In Sect. 2.1, we have made use of the fact that the potential is real-valued to show that the norm of any wavefunction is conserved. The fact that the total potential is now complex leads to a loss of norm. This may, however, be not as disturbing as the reentrance phenomenon, especially in situations where the wavepacket would just move on in "free-space" as e.g., in a scattering situation after the scattering event is over.

The choice of the functional form of $f(x)$ in (2.169) is crucial. It turns out that the potential has to rise smoothly and rather slowly in order that there do not occur unphysical reflections of the wavefunction induced by the negative imaginary potential. A detailed study of several different functional forms of the imaginary potential can be found in [33].

2.3.3 Alternative Methods of Time-Evolution

In the material presented so far we have dealt both, with the solution of problem (a) as well as problem (b). In the following some alternative ways of treating the time-evolution, i.e., problem (b) shall be reviewed.

Method of Finite Differences

The discretization of the time-derivative in (2.1) with the help of the first-order formula

$$|\dot{\Psi}(t)\rangle \approx \frac{|\Psi(t+\Delta t)\rangle - |\Psi(t)\rangle}{\Delta t} \tag{2.170}$$

leads to an *explicit* numerical method. This means that the wavefunction at a later time is explicitly given by the wavefunction at the earlier time. Unfortunately, however, this so-called Euler method is numerically instable.

An at least conditionally stable method can be constructed by application of the second-order formula

$$|\dot{\Psi}(t)\rangle \approx \frac{|\Psi(t+\Delta t)\rangle - |\Psi(t-\Delta t)\rangle}{2\Delta t} \tag{2.171}$$

for the time-derivative. The corresponding method is referred to as second-order differencing (SOD) and has been used for the solution of the time-dependent Schrödinger equation by Askar and Cakmak [34]. The method can be shown to be energy and norm conserving. The condition under which it is stable can be derived by considering the eigenvalues of the propagation matrix that appears by using the discrete form of the time-derivative

$$\begin{pmatrix} |\Psi_{n+1}\rangle \\ |\Psi_n\rangle \end{pmatrix} = \begin{pmatrix} \hat{1} - 4\hat{H}^2\Delta t^2/\hbar^2 & -2i\hat{H}\Delta t/\hbar \\ -2i\hat{H}\Delta t/\hbar & \hat{1} \end{pmatrix} \begin{pmatrix} |\Psi_{n-1}\rangle \\ |\Psi_{n-2}\rangle \end{pmatrix}. \tag{2.172}$$

The eigenvalues of the matrix are (replacing $\hat{H}$ by E)

$$\lambda_{1,2} = 1 - 2E^2\Delta t^2/\hbar^2 \pm \frac{2E\Delta t}{\hbar}\sqrt{\frac{E^2\Delta t^2}{\hbar^2} - 1}. \tag{2.173}$$

The discrete mapping is norm conserving due to $\lambda_1\lambda_2 = 1$. For stability, the radicant in the expression given above has to be negative, such that the eigenvalues become complex. Otherwise after the n-th iteration, an exponential increase of numerical instabilities would occur. Thus, for stability $\Delta t < \hbar/E_{max}$ has to hold, where E_{max} is the largest eigenvalue of $\hat{H}$ taking part in the dynamics [35]. A slightly different look at the second-order differencing method method is taken in Exercise 2.12.

Exercise 2.12 *Show that in the second-order differencing method the following holds if $\hat{H}$ is Hermitian*

(a) $\mathrm{Re}\langle\Psi(t-\Delta t)|\Psi(t)\rangle = \mathrm{Re}\langle\Psi(t)|\Psi(t+\Delta t)\rangle = \mathrm{const}$
(b) $\mathrm{Re}\langle\Psi(t-\Delta t)|\hat{H}|\Psi(t)\rangle = \mathrm{Re}\langle\Psi(t)|\hat{H}|\Psi(t+\Delta t)\rangle = \mathrm{const}$
(c) Interpret the results gained above.
(d) Consider the time-evolution of an eigenstate ψ of the Hamiltonian with eigenvalue λ and derive a criterion for the maximally allowed time step Δt. Hint: Insert the exact time-evolution into the SOD scheme and distinguish the exact eigenvalue from the approximate λ_{app} due to SOD time evolution.

Crank–Nicholson Method

An alternative possibility to circumvent the problem of instability of the Euler method is given by the Crank–Nichelson procedure. Here the first-order formula

$$\hat{U}(\Delta t) \approx \hat{1} - i\hat{H}\Delta t/\hbar \tag{2.174}$$

representing the infinitesimal time-evolution operator is used forward as well as backward in time

$$|\Psi_{n+1}\rangle = \hat{U}(\Delta t)|\Psi_n\rangle \tag{2.175}$$

$$|\Psi_{n-1}\rangle = \hat{U}(-\Delta t)|\Psi_n\rangle. \tag{2.176}$$

To make progress one resolves both equations for $|\Psi_n\rangle$ by multiplying with the respective inverse operators. Equating the gained expressions yields

$$(\hat{1} + i\hat{H}\Delta t/\hbar)|\Psi_{n+1}\rangle = (\hat{1} - i\hat{H}\Delta t/\hbar)|\Psi_{n-1}\rangle. \tag{2.177}$$

The procedure now is an implicit one, that is stable as well as norm conserving.

Due to its implicit nature the method requires a matrix inversion and formally leads to the prescription (also referred to as Cayley approximation)

$$|\Psi_n\rangle = \frac{\hat{1} - i\hat{H}\Delta t/(2\hbar)}{\hat{1} + i\hat{H}\Delta t/(2\hbar)}|\Psi_{n-1}\rangle \tag{2.178}$$

for the calculation of the propagated wavefunction.

Polynomial Methods

The idea behind polynomial methods is the expansion of the time-evolution operator in terms of polynomials, according to

$$e^{-i\hat{H}t/\hbar} = \sum_n a_n P_n(\hat{H}). \tag{2.179}$$

Two different approaches are commonly used:

- In the Chebyshev method the polynomials are fixed to be the complex valued Chebyshev ones. A first application to the problem of wavefunction propagation has been presented by Tal-Ezer and Kosloff [36]. These authors have shown that the approach is up to six times more efficient than the SOD method, presented above. It allows for evolution over relatively long time steps. Drawbacks are that intermediate time information is not available and even worse, that time-dependent Hamiltonians cannot be treated.
- In contrast to the first approach, in the Lanczos method the polynomials are not fixed but are generated in the course of the propagation. A very profound introduction to the commonly applied short iterative Lanczos method can be found in [16].

2.3.4 Semiclassical Initial Value Representations

As the final prerequisite before we deal with the physics of laser-matter interaction, a reformulation of the semiclassical van Vleck–Gutzwiller propagator presented in Sect. 2.2.2 shall be discussed. We have already mentioned that the VVG method is based on the solution of classical boundary value (or root search) problems which makes it hard to implement. A much more user friendly approach would be based on classical initial value solutions and is therefore termed initial value representation. We therefore start the discussion with a short introduction to commonly used symplectic integration procedures for the solution of the classical dynamics.

Symplectic Integration

Positions and momenta of a classical Hamiltonian system with N degrees of freedom obey the equations of motion

$$\dot{q}_n = \frac{\partial H(p_n, q_n, t)}{\partial p_n} \tag{2.180}$$

$$\dot{p}_n = -\frac{\partial H(p_n, q_n, t)}{\partial q_n}. \tag{2.181}$$

Using the Poisson bracket

$$\{a, b\} = \sum_{n=1}^{N} \left(\frac{\partial a}{\partial q_n} \frac{\partial b}{\partial p_n} - \frac{\partial a}{\partial p_n} \frac{\partial b}{\partial q_n} \right) \tag{2.182}$$

the equations above can also be cast into an equation for the phase space vector $\boldsymbol{\eta} = (\boldsymbol{q}, \boldsymbol{p})$

$$\dot{\boldsymbol{\eta}} = -\{H, \boldsymbol{\eta}\} =: -\hat{H}\boldsymbol{\eta}. \tag{2.183}$$

Although, we are dealing with classical mechanics an operator, $\hat{H}$, appears here. In the present subsection this operator stands for the application of the Poisson bracket with the Hamilton function.

Formally, the equation above can be integrated over a small time step, yielding

$$\eta(t + \Delta t) = \exp\{-\Delta t \hat{H}\}\eta(t). \tag{2.184}$$

Now we can again use the split-operator method, i.e., an "effective Hamiltonian" can be introduced according to

$$\exp\{-\Delta t \hat{H}_{\text{eff}}\} := \exp\{-\Delta t \hat{T}_{\text{k}}\} \exp\{-\Delta t \hat{V}\}. \tag{2.185}$$

This effective operator is only an approximation to the true one. However, the total phase space volume is conserved, i.e., Liouville's theorem holds also for the approximate dynamics [37].

In the numerics, the splitting of the Hamiltonian into the kinetic and the potential part means that first one solves a problem in which only $\hat{V}$ operates, i.e., the momentum is altered due to the effect of the potential; this is the so-called *kick* step. With the new momentum the position is then changed. This is the so-called *drift* step. For very short times (expand the exponential to first order) one gets

$$p^1 = p^0 + \Delta t F_{q=q^0}$$
$$q^1 = q^0 + \Delta t G_{p=p^1},$$

where the superscript denotes the iteration step and the abbreviations $G = \partial T_{\text{k}}/\partial p$ and $F = -\partial V/\partial q$ have been used. This procedure is a variant of the symplectic Euler method. It performs much better than the "standard" Euler method in which in the second line the old momentum p^0 is used.

Analogously to the discussion of the split-operator procedure in quantum mechanics, a split-operator procedure of higher order can be used. By employing

$$\exp\{-\Delta t \hat{H}_{\text{eff}}\} := \exp\{-\Delta t \hat{V}/2\} \exp\{-\Delta t \hat{T}_{\text{k}}\} \exp\{-\Delta t \hat{V}/2\}, \tag{2.186}$$

the so-called leap frog method

$$q^1 = q^0 + \Delta t/2 G(p = p^0),$$
$$p^2 = p^0 + \Delta t F(q = q^1),$$
$$q^2 = q^1 + \Delta t/2 G(p = p^2),$$

arises. In general, any symplectic integration scheme (where the kick step comes first) can be cast into the form

$$\begin{aligned} \text{do} \quad &k = 1, M \\ &p^k = p^{k-1} + b_k \Delta t F(q^{k-1}) \\ &q^k = q^{k-1} + a_k \Delta t G(p^k) \\ \text{enddo} \end{aligned}$$

Table 2.3. Coefficients for some symplectic integration methods of increasing order

Ruth's leap frog	$a_1 = 1/2$	$b_1 = 0$
	$a_2 = 1/2$	$b_2 = 1$
Fourth-order Gray [38]	$a_1 = (1 - \sqrt{1/3})/2$	$b_1 = 0$
	$a_2 = \sqrt{1/3}$	$b_2 = (1/2 + \sqrt{1/3})/2$
	$a_3 = -a_2$	$b_3 = 1/2$
	$a_4 = (1 + \sqrt{1/3})/2$	$b_4 = (1/2 - \sqrt{1/3})/2$
Sixth-order Yoshida [39]	$a_1 = 0.78451361047756$	$b_1 = 0.39225680523878$
	$a_2 = 0.23557321335936$	$b_2 = 0.51004341191846$
	$a_3 = -1.1776799841789$	$b_3 = -0.47105338540976$
	$a_4 = 1.3151863206839$	$b_4 = 0.068753168252520$
	$a_5 = a_3$	$b_5 = b_4$
	$a_6 = a_2$	$b_6 = b_3$
	$a_7 = a_1$	$b_7 = b_2$
	$a_8 = 0$	$b_8 = b_1$

Coefficients of some different symplectic methods are gathered in Table 2.3. Additional coefficients can be found in [37]. To keep the numerical effort of force calculation rather low, it is desirable to have as many b-coefficients equal to zero as possible.

Coherent States

Having set the stage with the discussion of the solution of the classical equations of motion, we now come to the central ingredient for the reformulation of the semiclassical propagator expression. These are the so-called coherent states which are discussed in detail in the text book of Louisell [40] and in Heller's Les Houches lecture notes [41]. In Dirac notation they are given by

$$|z\rangle = e^{-1/2|z|^2} e^{z\hat{a}^\dagger} |0\rangle, \qquad (2.187)$$

where $|0\rangle$ is the ground state of N uncoupled harmonic oscillators of mass m and frequency ω_e. Furthermore, we have used the multi-dimensional analog of (2.138)

$$\hat{a}^\dagger = \frac{1}{\sqrt{2}} \left(\frac{\hat{q}}{b} - i\frac{\hat{p}}{c} \right), \qquad (2.188)$$

for the vector of creation operators with $b = \sqrt{\hbar/m\omega_e}, c = \sqrt{\hbar m\omega_e}$ and

$$z = \frac{1}{\sqrt{2}} \left(\frac{q}{b} + i\frac{p}{c} \right), \qquad (2.189)$$

with the expectation values q and p of the operators $\hat{q}$ and $\hat{p}$.

In position representation, the coherent states are N-dimensional Gaussian wavepackets of the form

$$\langle \boldsymbol{x} | \boldsymbol{z} \rangle = \left(\frac{1}{\pi b^2} \right)^{N/4} \exp \left\{ -\frac{1}{2b^2} (\boldsymbol{x} - \boldsymbol{q})^2 + \frac{\mathrm{i}}{bc} \boldsymbol{p} \cdot \left(\boldsymbol{x} - \frac{\boldsymbol{q}}{2} \right) \right\}. \quad (2.190)$$

As proved, e.g., in [40] coherent states form an (over-) complete set of basis states and can be used to represent unity according to

$$\hat{1} = \int \frac{\mathrm{d}^{2N} z}{\pi^N} | \boldsymbol{z} \rangle \langle \boldsymbol{z} | = \int \frac{\mathrm{d}^N p \, \mathrm{d}^N q}{(2\pi\hbar)^N} | \boldsymbol{z} \rangle \langle \boldsymbol{z} |. \quad (2.191)$$

The basis for the reformulation of the semiclassical propagator is the matrix element of the time-evolution operator[13] between coherent states

$$K(\boldsymbol{z}_f, t; \boldsymbol{z}_i, 0) \equiv \langle \boldsymbol{z}_f | \mathrm{e}^{-\mathrm{i}\hat{H}t/\hbar} | \boldsymbol{z}_i \rangle. \quad (2.192)$$

The semiclassical approximation for this object can be performed quite analogously to the derivation of the van Vleck–Gutzwiller propagator by starting from the appropriate path integral [42]. However, it turns out that the final expression contains a classical over-determination problem due to the fact that not only the position is fixed at the initial and the final time but also the momentum! This problem is solved by the complexification of phase space. We will not dwell on that rather involved topic any longer. Fortunately the over-determination problem will be resolved rather elegantly in the following.

Herman–Kluk Propagator

The next step to make progress is to consider the time-evolution operator in position representation. It can be expressed via the coherent state propagator by inserting unity in terms of coherent states twice according to

$$
\begin{aligned}
K(\boldsymbol{x}_f, t; \boldsymbol{x}_i, 0) &= \langle \boldsymbol{x}_f | \mathrm{e}^{-\mathrm{i}\hat{H}t/\hbar} | \boldsymbol{x}_i \rangle \\
&= \int \frac{\mathrm{d}^{2N} z_f}{\pi^N} \int \frac{\mathrm{d}^{2N} z_i}{\pi^N} \langle \boldsymbol{x}_f | \boldsymbol{z}_f \rangle \langle \boldsymbol{z}_f | \mathrm{e}^{-\mathrm{i}\hat{H}t/\hbar} | \boldsymbol{z}_i \rangle \langle \boldsymbol{z}_i | \boldsymbol{x}_i \rangle.
\end{aligned}
$$

If we now replace the coherent state matrix element of the propagator by its semiclassical approximation and perform the final phase space integration in the stationary phase approximation, the over-determination problem is resolved and the semiclassical propagator is reformulated in terms of real classical initial value solutions [43]. This procedure yields

$$
\begin{aligned}
K^{\mathrm{HK}}(\boldsymbol{x}_f, t; \boldsymbol{x}_i, 0) &\equiv \int \frac{\mathrm{d}^N p_i \, \mathrm{d}^N q_i}{(2\pi\hbar)^N} \langle \boldsymbol{x}_f | \tilde{\boldsymbol{z}}_t \rangle R(\boldsymbol{p}_i, \boldsymbol{q}_i, t) \\
&\qquad \exp \left\{ \frac{\mathrm{i}}{\hbar} S(\boldsymbol{p}_i, \boldsymbol{q}_i, t) \right\} \langle \tilde{\boldsymbol{z}}_i | \boldsymbol{x}_i \rangle, \quad (2.193)
\end{aligned}
$$

---

[13] For notational convenience, we assume the Hamiltonian to be time-independent; the following results are also valid in the general case of a time-dependent Hamiltonian, however.

which is the so-called Herman–Kluk propagator. By a different reasoning it has first been devised by Herman and Kluk [44]. Definitions that are used in the expression above are the classical action *functional*, that depends on the initial phase space variables and for this reason is written (and denoted) as a *function* here according to

$$S(\boldsymbol{p}_i, \boldsymbol{q}_i, t) \equiv \int_0^t \mathrm{d}t' \left[\boldsymbol{p}_{t'} \cdot \dot{\boldsymbol{q}}_{t'} - H \right] \qquad (2.194)$$

Furthermore

$$R(\boldsymbol{p}_i, \boldsymbol{q}_i, t) \equiv \left| \frac{1}{2} \left(\mathbf{m}_{11} + \mathbf{m}_{22} - \mathrm{i}\hbar\gamma\mathbf{m}_{21} - \frac{1}{\mathrm{i}\hbar\gamma}\mathbf{m}_{12} \right) \right|^{1/2} \qquad (2.195)$$

with $\gamma = m\omega_\mathrm{e}/\hbar$ denotes the Herman–Kluk determinantal prefactor, which contains classical stability (monodromy) block-matrices $\mathbf{m}_{ij}$. They are solutions to the linearized Hamilton equations and for reference they are defined in Appendix 2.C.

The Gaussian wavepackets in (2.193) have a slightly different phase convention than the ones of (2.190). For this reason a new symbol with a tilde

$$\langle \boldsymbol{x} | \tilde{\boldsymbol{z}} \rangle = \left(\frac{\gamma}{\pi} \right)^{N/4} \exp \left\{ -\frac{\gamma}{2}(\boldsymbol{x} - \boldsymbol{q})^2 + \frac{\mathrm{i}}{\hbar} \boldsymbol{p} \cdot (\boldsymbol{x} - \boldsymbol{q}) \right\}, \qquad (2.196)$$

has been introduced. Due to the fact that the width parameters of the Gaussians are constant, the Herman–Kluk (HK) propagator is also referred to as the "Frozen Gaussian Approximation". To complete the explanation of all abbreviations, the centers of the final Gaussians in phase space are $\{\boldsymbol{p}_t(\boldsymbol{p}_i, \boldsymbol{q}_i), \boldsymbol{q}_t(\boldsymbol{p}_i, \boldsymbol{q}_i)\}$, which are the solutions of the classical Hamilton equations.

It is worthwhile to note that the expression (2.195) does not exhibit singularities at caustics. It has actually been proven that the Herman–Kluk method is a uniform semiclassical method [45]. For the numerics it is furthermore important that the square root in the prefactor has to be taken in such a fashion that the result is continuous as a function of time. This is reminiscent of the Maslov phase in the van Vleck–Gutzwiller expression (2.70), which does not have to be calculated explicitly, however.

A final remark on the connection between the two semiclassical expressions for the propagator, which we have discussed so far, shall be made. After performing the integration over the initial phase space variables in (2.193) in the stationary phase approximation the van Vleck–Gutzwiller expression will emerge. One can also turn around that reasoning and derive the Herman–Kluk prefactor by demanding that the SPA applied to the phase space integral yields the van Vleck–Gutzwiller expression [46]. For the derivation of a more general form of the prefactor in this way, see e.g., [47]. An even simpler way to derive the VVG propagator from the Herman–Kluk expression by taking the limit $\gamma \to \infty$ is explicitly given in Appendix 2.D.

Semiclassical Propagation of Gaussian Wavepackets

The pure HK propagator is a clumsy object due to the need to integrate over all of phase space. Fortunately, however, in the focus of our interest will not be the bare propagator but its application to an initial Gaussian wavepacket. Let us therefore consider the mixed matrix element

$$K(\boldsymbol{x}_f, t; \tilde{\boldsymbol{z}}_\alpha, 0) \equiv \langle \boldsymbol{x}_f | e^{-i\hat{H}t/\hbar} | \tilde{\boldsymbol{z}}_\alpha \rangle = \int \mathrm{d}^N x_i K(\boldsymbol{x}_f, t; \boldsymbol{x}_i, 0) \langle \boldsymbol{x}_i | \tilde{\boldsymbol{z}}_\alpha \rangle, \quad (2.197)$$

of the time-evolution operator, where $K(\boldsymbol{x}_f, t; \boldsymbol{x}_i, 0)$ shall be replaced by the HK approximation of (2.193).

The Gaussian to be propagated $\langle \boldsymbol{x}_i | \tilde{\boldsymbol{z}}_\alpha \rangle$ shall be determined by its phase space center $(\boldsymbol{q}_\alpha, \boldsymbol{p}_\alpha)$ and shall have the same width parameter γ as the coherent state basis functions. The calculation of the overlap in (2.197) can be done analytically and yields the simple result

$$\langle \tilde{\boldsymbol{z}}_i | \tilde{\boldsymbol{z}}_\alpha \rangle = \exp \left\{ -\frac{\gamma}{4}(\boldsymbol{q}_i - \boldsymbol{q}_\alpha)^2 + \frac{i}{2\hbar}(\boldsymbol{q}_i - \boldsymbol{q}_\alpha) \cdot (\boldsymbol{p}_i + \boldsymbol{p}_\alpha) \right.$$
$$\left. - \frac{1}{4\gamma\hbar^2}(\boldsymbol{p}_i - \boldsymbol{p}_\alpha)^2 \right\}. \quad (2.198)$$

In (2.197), the integration over initial phase space still has to be done. It is, however, much more "user friendly" than in the case of the bare propagator, due to the fact that the overlap just calculated is effectively cutting off the integrand too far away from the initial center in phase space. In numerical applications the phase space integration is often performed by using Monte Carlo methods [48]. Pictorially, the application of the Herman–Kluk propagator to a Gaussian can be represented as shown in Fig. 2.6.

How is all that related to the (thawed) Gaussian wavepacket dynamics (GWD) of Heller that we have used in Sect. 2.1.4? There the Gaussian wavepacket has been propagated using its center trajectory alone. The GWD therefore is much more crude than the HK method! There should be a way to derive GWD from the more general expression, however. This is indeed the case. To this end one has to expand the exponent in the integral over initial phase space around the center $(\boldsymbol{p}_\alpha, \boldsymbol{q}_\alpha)$ of the initial Gaussian up to second order. The integration is then a Gaussian integration and can be performed analytically.[14] One finally gets [47]

$$K^{\mathrm{GWD}}(\boldsymbol{x}_f, t; \tilde{\boldsymbol{z}}_\alpha, 0) \equiv \left(\frac{\gamma}{\pi}\right)^{N/4} |(\mathbf{m}_{22} + i\hbar\gamma\mathbf{m}_{21})|^{-1/2}$$
$$\exp \left\{ -\frac{1}{2}(\boldsymbol{x}_f - \boldsymbol{q}_{\alpha t}) \cdot \gamma_t (\boldsymbol{x}_f - \boldsymbol{q}_{\alpha t}) \right.$$
$$\left. + \frac{i}{\hbar}\boldsymbol{p}_{\alpha t} \cdot (\boldsymbol{x}_f - \boldsymbol{q}_{\alpha t}) + \frac{i}{\hbar}S \right\}, \quad (2.199)$$

---

[14] Please note that this procedure is much more crude and approximative than a stationary phase approximation.

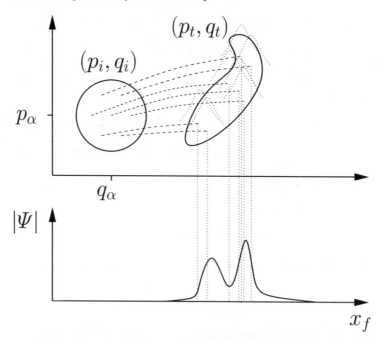

Fig. 2.6. Pictorial representation of the semiclassical initial value procedure to propagate a Gaussian wavepacket as a Herman and Kluk in one dimension: the propagated wavepacket $\langle x_f|\Psi\rangle$ is a weighted sum over many Gaussians. The weights are the product of a prefactor times a complex exponent $R\exp\{iS/\hbar\}\langle\tilde{z}_i|\tilde{z}_\alpha\rangle$; adapted from [49]

with the time-dependent $N\times N$ width parameter matrix

$$\gamma_t = \gamma\left(\mathbf{m}_{11} + \frac{1}{i\gamma\hbar}\mathbf{m}_{12}\right)\left(\mathbf{m}_{22} + i\gamma\hbar\mathbf{m}_{21}\right)^{-1}. \qquad (2.200)$$

The width of the single Gaussian can thus change in the course of time, in contrast to the widths of the many Gaussians in the case of the HK propagator. This is the reason why the more simple single Gaussian method is called "thawed" Gaussian wavepacket dynamics and the more complex multiple Gaussian HK method is called "frozen" Gaussian wavepacket dynamics. Finally, it is worthwhile to check that the time-dependent width parameter $\alpha_t = \gamma_t/2$ fulfills the nonlinear Riccati differential equation (2.45). To this end, in Exercise 2.13 the equations of motion of the stability matrix elements given in Appendix 2.C should be used.

Exercise 2.13 *Show that the nonlinear differential equation*

$$\dot{\alpha}_t = -\frac{2i\hbar}{m}\alpha_t^2 + \frac{i}{2\hbar}V''$$

is fulfilled by the width parameter

$$\alpha_t = \alpha_0 \frac{m_{11} + \frac{1}{2i\alpha_0\hbar}m_{12}}{m_{22} + 2i\alpha_0\hbar m_{21}},$$

where m_{ij} are elements of the stability matrix.

The different level of accuracy of the two approximations is illustrated in Fig. 2.7, where a comparison of the multi-trajectory HK method and the single trajectory GWD are contrasted with exact numerical results gained by using split-operator FFT. The displayed quantity is the auto-correlation function

$$c(t) \equiv \langle\Psi(0)|\Psi(t)\rangle \qquad (2.201)$$

of an initial wavepacket in a Morse potential with dimensionless Hamiltonian

$$\hat{H} = \frac{\hat{p}^2}{2} + D(1 - \exp\{-\lambda x\})^2, \qquad (2.202)$$

with $D = 30, \lambda = 0.08$.[15] GWD describes the envelope of the quantum curve rather well. The fine oscillations, that are captured almost perfectly by the multiple trajectory method are not described at all, however.

2.A The Royal Road to the Path Integral

An elegant proof of the representation of the path integral as an infinite dimensional Riemann integral can be performed by using the Weyl transformation [50]

$$A(p,q) = \int du e^{iqu/\hbar}\langle p + u/2|\hat{A}|p - u/2\rangle, \qquad (2.203)$$

that transforms an operator into a phase space function. The inverse transformation is

$$\hat{A} = \frac{1}{h}\int dp dq A(p,q)\hat{\Delta}(p,q) \qquad (2.204)$$

$$\hat{\Delta} = \int dv e^{ipv/\hbar}|q + v/2\rangle\langle q - v/2|. \qquad (2.205)$$

Using the time-slicing procedure introduced in Sect. 2.2.1, the propagator can be represented by a product of short-time propagators according to

$$K(x_f, t; x_i, 0) = \int dq_1 \ldots \int dq_{N-1} \prod_{k=0}^{N-1}\langle q_{k+1}|e^{-\frac{i}{\hbar}\hat{H}(\hat{p},\hat{q})\Delta t}|q_k\rangle, \qquad (2.206)$$

---

[15] In Sect. 5.1.2 the physical background of the Morse oscillator will be elucidated.

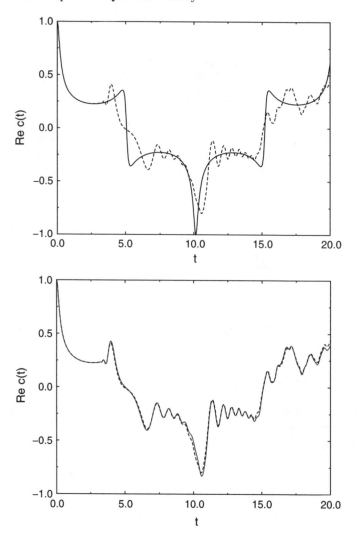

Fig. 2.7. Different trajectory-based and fully quantum mechanical auto-correlation function of a Gaussian wavepacket with dimensionless parameters $q_\alpha, p_\alpha = 0$ und $\gamma = 12$ in a Morse potential. *Upper panel:* thawed GWD (*solid line*) versus exact quantum mechanics (*dashed line*); *lower panel:* HK (*solid line*) versus exact quantum mechanics (*dashed line*)

where $q_N = x_f$ and $q_0 = x_i$ and, where we have switched from x to q notation for reasons of convenience. Using the Weyl transform of the Hamiltonian

$$H(p, q) = \frac{p^2}{2m} + V(q) \tag{2.207}$$

and (2.204, 2.205), the short-time propagators can be written as an integral

$$\langle q_{k+1}| \exp\left\{-\frac{i}{\hbar}\hat{H}(\hat{p},\hat{q})\Delta t\right\}|q_k\rangle$$

$$= \frac{1}{h}\langle q_{k+1}| \int \mathrm{d}p\mathrm{d}q \exp\left\{-\frac{i}{\hbar}H(p,q)\Delta t\right\} \int \mathrm{d}v \mathrm{e}^{ipv/\hbar}|q+v/2\rangle\langle q-v/2|q_k\rangle$$

$$= \frac{1}{h}\int \mathrm{d}p\,\mathrm{d}q \exp\left\{-\frac{i}{\hbar}H(p,q)\Delta t\right\} \mathrm{e}^{ip(q_{k+1}-q_k)/\hbar}\delta\left\{\frac{q_{k+1}+q_k}{2}-q\right\}$$

$$= \frac{1}{h}\int \mathrm{d}p\exp\left\{\frac{i}{\hbar}\left[p\frac{q_{k+1}-q_k}{\Delta t}-H\left(p,\frac{q_{k+1}+q_k}{2}\right)\right]\Delta t\right\}.$$

Inserting this expression into (2.206), the phase-space integral form of the propagator [50]

$$K(x_f,t;x_i,0) = \int \frac{\mathrm{d}p_0\ldots\mathrm{d}p_{N-1}\mathrm{d}q_1\ldots\mathrm{d}q_{N-1}}{(2\pi\hbar)^N}$$

$$\exp\left\{\frac{i}{\hbar}\sum_{k=0}^{N-1}\left[p_k\left(\frac{q_{k+1}-q_k}{\Delta t}\right)-H\left(p_k,\frac{q_{k+1}+q_k}{2}\right)\right]\Delta t\right\},$$

can be derived. Please note that the number of p-integrations is higher by one than the number of q-integrations. Furthermore, the expression above does not contain operators any more!

Due to the quadratic form of the Weyl transform (2.207) in p, the p integrals are Gaussian, however, and can be done exactly analytically. This leads to the expression (analogous to (2.53))

$$K(x_f,t;x_i,0) = \lim_{N\to\infty}\left[\frac{m}{2\pi i\hbar\Delta t}\right]^{N/2}\int \prod_{k=1}^{N-1}\mathrm{d}q_k$$

$$\exp\left\{\frac{i}{\hbar}\sum_{k=0}^{N-1}\left[\frac{(q_{k+1}-q_k)^2}{2\Delta t}-V\left(\frac{q_{k+1}+q_k}{2}\right)\Delta t\right]\right\},$$

where for $\Delta t \to 0$ the classical action appears in the exponent. Furthermore, the normalization constant (or measure factor) that was still undetermined in (2.53) is given by

$$B_N = \left[\frac{m}{2\pi i\hbar\Delta t}\right]^{N/2}. \tag{2.208}$$

2.B Variational Calculus

In general the variation of a functional is defined via

$$\delta\Phi \equiv \Phi[h+\delta h]-\Phi[h] = \int \mathrm{d}x\frac{\delta\Phi}{\delta h(x)}\delta h(x). \tag{2.209}$$

For the specific cases

$$\Phi_1[h] = \int_a^b dx \, h(x) f(x) \tag{2.210}$$

$$\Phi_2[h] = \int_a^b dx \, F(x, h(x)) \tag{2.211}$$

$$\Phi_3[h] = \int_a^b dx \, F\left(x, h(x), \frac{dh(x)}{dx}\right) \tag{2.212}$$

from the definition given above

$$\frac{\delta\Phi_1}{\delta h(x)} = f(x), \tag{2.213}$$

$$\frac{\delta\Phi_2}{\delta h(x)} = \frac{\partial F}{\partial h}, \tag{2.214}$$

$$\frac{\delta\Phi_3}{\delta h(x)} = \frac{\partial F}{\partial h} - \frac{d}{dx}\frac{\partial F}{\partial h'} \tag{2.215}$$

can be deduced for the functional derivative, and the variation is reduced to the calculation of well-known partial derivatives.

To perform the stationary phase approximation to the path integral, the second variation of the classical action functional S at the classical path is needed. To calculate this quantity, we are using the first equation in (2.209) and are considering the classical path $x_{cl}(t)$ and deviations η that vanish at the initial and final time $t' = 0, t$. The first variation of the action is then given by

$$\begin{aligned}
\delta S[x_{cl}] &= S[x_{cl} + \eta] - S[x_{cl}] \\
&= \int_0^t dt' \left\{ \frac{m}{2}\left[\frac{d}{dt'}(x_{cl} + \eta)\right]^2 - V(x_{cl} + \eta) \right\} - S[x_{cl}] \\
&= \int_0^t dt' \left\{ \frac{m}{2}\dot{x}_{cl}^2 + m\dot{x}_{cl}\dot{\eta} - V(x_{cl}) - V'(x_{cl})\eta \right\} - S[x_{cl}] \\
&= \int_0^t dt' \left\{ m\dot{x}_{cl}\dot{\eta} - V'(x_{cl})\eta \right\} \\
&= \int_0^t dt' \left\{ -m\ddot{x}_{cl}\eta - V'(x_{cl})\eta \right\} + m\dot{x}_{cl}\eta\big|_0^t \\
&\equiv 0. \tag{2.216}
\end{aligned}$$

The vanishing of the first variation of the action is Hamilton's principle of classical mechanics and in the simple 1d case here it leads to Newton's equation

$$m\ddot{x}_{cl} + V'(x_{cl}) = 0 \tag{2.217}$$

for the classical path, which we could have concluded directly from (2.215). Up to second order, we get

$$S = S[x_{\text{cl}}] + \frac{1}{2}\delta^2 S[x_{\text{cl}}] \tag{2.218}$$

with the second variation

$$
\begin{aligned}
\delta^2 S[x_{\text{cl}}] &= \delta S[x_{\text{cl}} + \eta] - \delta S[x_{\text{cl}}] \\
&= \int_0^t \mathrm{d}t' \left\{ -m\frac{\mathrm{d}^2}{\mathrm{d}t^2}(x_{\text{cl}} + \eta) - V'(x_{\text{cl}} + \eta) \right\} \eta \\
&\quad - \int_0^t \mathrm{d}t' \left\{ -m\ddot{x}_{\text{cl}} - V'(x_{\text{cl}}) \right\} \eta \\
&= \int_0^t \mathrm{d}t' \left\{ -m\ddot{\eta} - V''(x_{\text{cl}})\eta \right\} \eta \\
&= \int_0^t \mathrm{d}t'\, \eta \hat{O} \eta,
\end{aligned}
\tag{2.219}
$$

with the operator $\hat{O}$ from (2.63) of Sect. 2.2.1.

2.C Stability Matrix

We have seen that the central ingredient of the prefactor of the semiclassical Herman-Kluk propagator is the stability (or monodromy) matrix defined by

$$
\mathbf{M} = \begin{pmatrix} \mathbf{m}_{11} & \mathbf{m}_{12} \\ \mathbf{m}_{21} & \mathbf{m}_{22} \end{pmatrix} = \begin{pmatrix} \frac{\partial \boldsymbol{p}_t}{\partial \boldsymbol{p}_i^{\mathrm{T}}} & \frac{\partial \boldsymbol{p}_t}{\partial \boldsymbol{q}_i^{\mathrm{T}}} \\ \frac{\partial \boldsymbol{q}_t}{\partial \boldsymbol{p}_i^{\mathrm{T}}} & \frac{\partial \boldsymbol{q}_t}{\partial \boldsymbol{q}_i^{\mathrm{T}}} \end{pmatrix}.
\tag{2.220}
$$

This matrix determines the time-evolution of small deviations in the initial conditions of a specific trajectory according to

$$
\begin{pmatrix} \delta \boldsymbol{p}_t \\ \delta \boldsymbol{q}_t \end{pmatrix} = \mathbf{M} \begin{pmatrix} \delta \boldsymbol{p}_i \\ \delta \boldsymbol{q}_i \end{pmatrix},
\tag{2.221}
$$

where $\delta \boldsymbol{p}_t = \tilde{\boldsymbol{p}}_t - \boldsymbol{p}_t$ and $\delta \boldsymbol{q}_t = \tilde{\boldsymbol{q}}_t - \boldsymbol{q}_t$. Pictorially this is represented for one spatial dimension in Fig. 2.8.

The equations of motion for the stability matrix can be gained by linearizing Hamilton's equations for the deviations. After Taylor expansion up to first order we get

$$
\dot{\delta \boldsymbol{p}_t} = -\frac{\partial H}{\partial \boldsymbol{q}}(\tilde{\boldsymbol{p}}_t, \tilde{\boldsymbol{q}}_t) + \frac{\partial H}{\partial \boldsymbol{q}}(\boldsymbol{p}_t, \boldsymbol{q}_t) = -\frac{\partial^2 H}{\partial \boldsymbol{q} \partial \boldsymbol{p}^{\mathrm{T}}} \delta \boldsymbol{p}_t - \frac{\partial^2 H}{\partial \boldsymbol{q} \partial \boldsymbol{q}^{\mathrm{T}}} \delta \boldsymbol{q}_t
\tag{2.222}
$$

$$
\dot{\delta \boldsymbol{q}_t} = \frac{\partial H}{\partial \boldsymbol{p}}(\tilde{\boldsymbol{p}}_t, \tilde{\boldsymbol{q}}_t) - \frac{\partial H}{\partial \boldsymbol{p}}(\boldsymbol{p}_t, \boldsymbol{q}_t) = \frac{\partial^2 H}{\partial \boldsymbol{p} \partial \boldsymbol{p}^{\mathrm{T}}} \delta \boldsymbol{p}_t + \frac{\partial^2 H}{\partial \boldsymbol{p} \partial \boldsymbol{q}^{\mathrm{T}}} \delta \boldsymbol{q}_t,
\tag{2.223}
$$

where the vectors are coloumn vectors and their transposes are row vectors. The linear differential equation of first order for the stability matrix thus reads

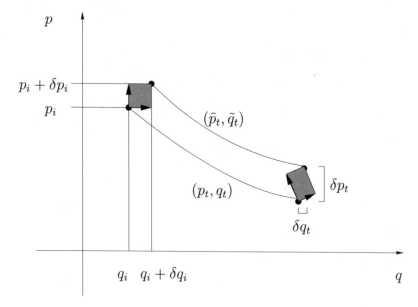

Fig. 2.8. Variation of initial conditions in phase space for one degree of freedom. The shaded area is constant in time (Liouville's theorem)

$$\frac{d}{dt}\mathbf{M} = -\mathbf{JHM},\qquad(2.224)$$

with the skew symmetric matrix

$$\mathbf{J} = \begin{pmatrix} \mathbf{0} & \mathbf{1} \\ -\mathbf{1} & \mathbf{0} \end{pmatrix}\qquad(2.225)$$

and the Hessian matrix of the Hamiltonian

$$\mathbf{H} = \begin{pmatrix} \frac{\partial^2 H}{\partial\mathbf{p}\partial\mathbf{p}^{\mathrm{T}}} & \frac{\partial^2 H}{\partial\mathbf{p}\partial\mathbf{q}^{\mathrm{T}}} \\ \frac{\partial^2 H}{\partial\mathbf{q}\partial\mathbf{p}^{\mathrm{T}}} & \frac{\partial^2 H}{\partial\mathbf{q}\partial\mathbf{q}^{\mathrm{T}}} \end{pmatrix}.\qquad(2.226)$$

The initial conditions follow directly from the definition (2.220) to be

$$\mathbf{M}(0) = \begin{pmatrix} \mathbf{m}_{11}(0) & \mathbf{m}_{12}(0) \\ \mathbf{m}_{21}(0) & \mathbf{m}_{22}(0) \end{pmatrix} = \begin{pmatrix} \mathbf{1} & \mathbf{0} \\ \mathbf{0} & \mathbf{1} \end{pmatrix}.\qquad(2.227)$$

Using the equations of motion (2.224) and the initial conditions, it can be shown that $\frac{d}{dt}\mathbf{M}^{\mathrm{T}}\mathbf{JM} = 0$ and therefore $\mathbf{M}^{\mathrm{T}}\mathbf{JM} = \mathbf{M}^{\mathrm{T}}(t=0)\mathbf{JM}(t=0) = \mathbf{J}$ for all times, i.e. $\mathbf{M}$ is a symplectic matrix. Written out more explicitly, the previous statement reads

$$\mathbf{m}_{22}^{\mathrm{T}}\mathbf{m}_{11} - \mathbf{m}_{12}^{\mathrm{T}}\mathbf{m}_{21} = 1 \qquad \forall t \qquad(2.228)$$

$$\mathbf{m}_{11}^{\mathrm{T}}\mathbf{m}_{21} - \mathbf{m}_{21}^{\mathrm{T}}\mathbf{m}_{11} = 0 \qquad \forall t \qquad(2.229)$$

$$\mathbf{m}_{22}^{\mathrm{T}}\mathbf{m}_{12} - \mathbf{m}_{12}^{\mathrm{T}}\mathbf{m}_{22} = 0 \qquad \forall t. \qquad(2.230)$$

The determinant of the stability matrix is equal to unity as can be shown by using the relations above[16]. This is equivalent to the conservation of phase space volume mentioned previously. In 1d one does need to take care of block matrix calculus and in addition Liouville's theorem can be checked by calculating the vector product of the vectors drawn in Fig. 2.8.

Finally it should be mentioned that the matrix element $\mathbf{m}_{21}$, up to a sign, is the inverse of the second derivative of the action appearing in the prefactor of the VVG propagator (2.70)

$$\mathbf{m}_{21} = -\left(\frac{\partial^2 S}{\partial \boldsymbol{q}_t \partial \boldsymbol{q}_i^{\mathrm{T}}}\right)^{-1}. \tag{2.231}$$

2.D From the HK- to the VVG-Propagator

The most straightforward way to derive the van Vleck-Gutzwiller from the Herman-Kluk propagator is by taking the limit $\gamma \to \infty$ in

$$K^{\mathrm{HK}}(\boldsymbol{x}_f, t; \boldsymbol{x}_i, 0) = \int \frac{\mathrm{d}^N p_i \mathrm{d}^N q_i}{(2\pi\hbar)^N} \langle \boldsymbol{x}_f | \tilde{\boldsymbol{z}}_t \rangle R(\boldsymbol{p}_i, \boldsymbol{q}_i, t)$$
$$\exp\left\{\frac{\mathrm{i}}{\hbar} S(\boldsymbol{p}_i, \boldsymbol{q}_i, t)\right\} \langle \tilde{\boldsymbol{z}}_i | \boldsymbol{x}_i \rangle, \tag{2.232}$$

which is reproduced here for convenience. The Gaussian wavepackets in that limit are "almost" δ-functions. δ-functions are normalized differently, however, and therefore in order to make use of their properties we note that in the limit $\gamma \to \infty$

$$\lim_{\gamma \to \infty} \left(\frac{4\pi}{\gamma}\right)^{-N/4} \langle \boldsymbol{x}_f | \tilde{\boldsymbol{z}}_t \rangle = \delta(\boldsymbol{x}_f - \boldsymbol{q}_t) \tag{2.233}$$

$$\lim_{\gamma \to \infty} \left(\frac{4\pi}{\gamma}\right)^{-N/4} \langle \tilde{\boldsymbol{z}}_i | \boldsymbol{x}_i \rangle = \delta(\boldsymbol{x}_i - \boldsymbol{q}_i). \tag{2.234}$$

The inverse of the γ-dependent factors in front of the Gaussians together with the prefactor (2.195) give

$$\lim_{\gamma \to \infty} \left(\frac{4\pi}{\gamma}\right)^{N/2} R(\boldsymbol{p}_i, \boldsymbol{q}_i, t) = (2\pi\hbar\mathrm{i})^{N/2}\sqrt{-|\mathbf{m}_{21}|}, \tag{2.235}$$

where the bars stand for taking the determinant of the stability submatrix. The $\boldsymbol{q}_i$ integration can be performed immediately allowing us to replace $\boldsymbol{q}_i$ by $\boldsymbol{x}_i$. For the $\boldsymbol{p}_i$ integration we use the relation

---

[16] Be careful to use the formula $\det \mathbf{M} = \det \mathbf{m_{22}} \det(\mathbf{m_{11}} - \mathbf{m_{12}}\mathbf{m_{22}^{-1}}\mathbf{m_{21}})$ valid for block matrices!

$$\delta[\boldsymbol{x}_f - \boldsymbol{q}_t(\boldsymbol{p}_i, \boldsymbol{x}_i)] = \sum_j \frac{1}{||\partial \boldsymbol{q}_t/\partial \boldsymbol{p}_i||} \delta(\boldsymbol{p}_i - \boldsymbol{p}_j) \qquad (2.236)$$

for δ-functions of functions of the integration variable. Here we have to sum over all momenta $\boldsymbol{p}_j$ leading to zeros of $\boldsymbol{x}_f - \boldsymbol{q}_t(\boldsymbol{p}_i, \boldsymbol{x}_i)$ and the double bars denote the absolute value of the determinant.

Due to $\partial \boldsymbol{q}_t/\partial \boldsymbol{p}_i = \mathbf{m}_{21}$ and with (2.231) from Appendix 2.C, we finally arrive at the N degree of freedom form of the VVG expression

$$K^{\mathrm{VVG}}(\boldsymbol{x}_f, t; \boldsymbol{x}_i, 0) = \left(\frac{\mathrm{i}}{2\pi\hbar}\right)^{\frac{N}{2}} \sum_j \sqrt{\det\left(\frac{\partial^2 S_j}{\partial \boldsymbol{x}_f \partial \boldsymbol{x}_i^{\mathrm{T}}}\right)} \exp\left\{\frac{\mathrm{i}}{\hbar} S_j(\boldsymbol{x}_f, \boldsymbol{x}_i)\right\}.$$

Notes and Further Reading

The very detailed book by Schleich [51] contains a lot of additional information on time-evolution operators and also the time-evolution of the density operator is discussed therein. More on time-dependent and energy-dependent Green's function can be found in Economou's book [52]. The extraction of spectral information from time-dependent quantum information goes back to work of Heller, which is reviewed in [53] and is covered also in the book by Tannor [16]. In both references also many additional references concerning Gaussian wavepacket dynamics can be found.

A lot of additional material on the path integral formulation of quantum mechanics is contained in the tutorial article by Ingold [54] and in the books by Feynman and Hibbs [11] and by Schulman [12]. The second book has more details on variational calculus and on exactly solvable path integrals (especially in the new Dover edition). The book by Reichl [55] contains chapters on semiclassical methods and on time-periodic systems, dealing with Floquet theory. These methods are then used in the context of quantum chaology. The book by Billing contains more material on semiclassics and on mixed quantum classical methods [56].

The book by Tannor [16] contains more information on methodological and numerical approaches to solve the time-dependent Schrödinger equation. Polynomial and DVR methods, which we have only touched upon shortly, are e.g., dealt with in detail there. A book that is fun to read, although it covers a seemingly dry topic is the classic "Numerical Recipes" [24]. Among many other things, more details on FFT can be found therein. A review of different semiclassical approximations based on Gaussian wavepackets is given in [57].

References

1. E. Schrödinger, Ann. Phys. (Leipzig) **79**, 361 (1926)
2. E. Schrödinger, Ann. Phys. (Leipzig) **81**, 109 (1926)

3. J.S. Briggs, J.M. Rost, Eur. Phys. J. D **10**, 311 (2000)
4. V.A. Mandelshtam, J. Chem. Phys. **108**, 9999 (1998)
5. E. Schrödinger, Die Naturwissenschaften **14**, 664 (1926)
6. E.J. Heller, J. Chem. Phys. **62**, 1544 (1975)
7. W. Kinzel, Phys. B. **51**, 1190 (1995)
8. F. Grossmann, J.M. Rost, W.P. Schleich, J. Phys. A **30**, L277 (1997)
9. R.P. Feynman, Rev. Mod. Phys. **20**, 367 (1948)
10. P.A.M. Dirac, *The Principles of Quantum Mechanics*, 4th edn. (Oxford, London, 1958)
11. R.P. Feynman, A.R. Hibbs, *Quantum Mechanics and Path Integrals* (McGraw-Hill, New York, 1965). For errata see e.g. http://www.oberlin.edu/physics/dyster/FeynmanHibbs/
12. L.S. Schulman, *Techniques and Applications of Path Integration* (Dover, Mineola, 2006)
13. J.H. van Vleck, Proc. Acad. Nat. Sci. USA **14**, 178 (1928)
14. M.C. Gutzwiller, J. Math. Phys. **8**, 1979 (1967)
15. S. Grossmann, *Funktionalanalysis II* (Akademic, Wiesbaden, 1977)
16. D.J. Tannor, *Introduction to Quantum Mechanics: A Time-dependent Perspective* (University Science Books, Sausalito, 2007)
17. W.R. Salzman, J. Chem. Phys. **85**, 4605 (1986)
18. M.H. Beck, A. Jäckle, G.A. Worth, H.D. Meyer, Phys. Rep. **324**, 1 (2000)
19. D. Kohen, F. Stillinger, J.C. Tully, J. Chem. Phys. **109**, 4713 (1998)
20. M. Tinkham, *Group Theory and Quantum Mechanics* (McGraw-Hill, New York, 1964)
21. H. Sambe, Phys. Rev. A **7**, 2203 (1973)
22. M. Kleber, Phys. Rep. **236**, 331 (1994)
23. M. Abramowitz, I.A. Stegun, *Handbook of Mathematical Functions* (Dover, New York, 1965)
24. W.H. Press, S.A. Teukolsky, W.T. Vetterling, B.P. Flannery, *Numerical Recipes in Fortran*, 2nd edn. (Cambridge University Press, Cambridge, 1992)
25. J.H. Shirley, Phys. Rev. **138**, B979 (1965)
26. J.C. Light, in *Time-Dependent Quantum Molecular Dynamics*, ed. by J. Broeckhove, L. Lathouwers (Plenum, New York, 1992), p. 185
27. U. Peskin, N. Moiseyev, Phys. Rev. A **49**, 3712 (1994)
28. W. Witschel, J. Phys. A **8**, 143 (1975)
29. J.A. Fleck, J.R. Morris, M.D. Feit, Appl. Phys. **10**, 129 (1976)
30. M.D. Feit, J.A. Fleck, A. Steiger, J. Comp. Phys. **47**, 412 (1983)
31. M. Braun, C. Meier, V. Engel, Comp. Phys. Commun. **93**, 152 (1996)
32. M. Frigo, S.G. Johnson, Proceedings of the IEEE **93**(2), 216 (2005). Special issue on "Program Generation, Optimization, and Platform Adaptation"
33. A. Vibok, G.G. Balint-Kurti, J. Phys. Chem **96**, 8712 (1992)
34. A. Askar, A.S. Cakmak, J. Chem. Phys. **68**, 2794 (1978)
35. C. Leforestier, R.H. Bisseling, C. Cerjan, M.D. Feit, R. Friesner, A. Guldberg, A. Hammerich, G. Jolicard, W. Karrlein, H.D. Meyer, N. Lipkin, O. Roncero, R. Kosloff, J. Comp. Phys. **94**, 59 (1991)
36. H. Tal-Ezer, R. Kosloff, J. Chem. Phys. **81**, 3967 (1984)
37. S.K. Gray, D.W. Noid, B.G. Sumpter, J. Chem. Phys. **101**, 4062 (1994)
38. M.L. Brewer, J.S. Hulme, D.E. Manolopoulos, J. Chem. Phys. **106**, 4832 (1997)
39. H. Yoshida, Phys. Lett. A **150**, 262 (1990)

40. W.H. Louisell, *Quantum Statistical Properties of Radiation* (Wiley, New York, 1990)
41. E.J. Heller, J. Chem. Phys. **94**, 2723 (1991)
42. J.R. Klauder, in *Random Media*, ed. by G. Papanicolauou (Springer, Berlin Heidelberg New York, 1987), p. 163
43. F. Grossmann, J. A. L. Xavier, Phys. Lett. A **243**, 243 (1998)
44. M.F. Herman, E. Kluk, Chem. Phys. **91**, 27 (1984)
45. K.G. Kay, Chem. Phys. **322**, 3 (2006)
46. K.G. Kay, J. Chem. Phys. **100**, 4377 (1994)
47. F. Grossmann, J. Chem. Phys. **125**, 014111 (2006)
48. E. Kluk, M.F. Herman, H.L. Davis, J. Chem. Phys. **84**, 326 (1986)
49. F. Grossmann, M. Herman, J. Phys. A **35**, 9489 (2002)
50. M. Mizrahi, J. Math. Phys. **16**, 2201 (1975)
51. W.P. Schleich, *Quantum Optics in Phase Space* (Wiley, Berlin, 2000)
52. E.N. Economou, *Green's Functions in Quantum Physics*, 3rd edn. (Springer, Berlin Heidelberg New York, 2006)
53. E.J. Heller, in *Chaos and Quantum Physics*, ed. by M.J. Giannoni, A. Voros, J. Zinn-Justin, Les Houches Session LII (Elsevier, Amsterdam, 1991), pp. 549–661
54. G.-L. Ingold, in *Coherent Evolution in Noisy Environments*, ed. by A. Buchleitner and K. Hornberger, Lecture Notes in Physics **611** (Springer, Berlin Heidelberg New York, 2002), pp. 1–53
55. L.E. Reichl, *The Transition to Chaos*, 2nd edn. (Springer, Berlin Heidelberg New York, 2004)
56. G.D. Billing, *The Quantum Classical Theory* (Oxford University Press, New York, 2003)
57. F. Grossmann, Comments At. Mol. Phys. **34**, 141 (1999)

Part II

Applications

3
_____

Field Matter Coupling and Two-Level Systems

With this chapter we start the applications part of this book by considering the interaction between lasers and matter. Lasers have already been discussed in Chap. 1. Therefore, we start immediately with the theoretical description of the coupling of a given classical light field, realized e.g. by a laser, to a quantum mechanical system.[1]

Due to their simplicity and the fact that they serve as paradigms for many phenomena observed in more complex systems, some analytically solvable two-level systems will be discussed in the remainder of this chapter.

3.1 Light Matter Interaction

The interaction of a single quantum particle with an electromagnetic field shall be considered in the following. Starting from the principle of minimal coupling and using several unitary transformations, some commonly used ways of setting up a field-driven Hamiltonian will be presented.

3.1.1 Minimal Coupling

The most straightforward route to the coupling of a charged particle with charge q to an electromagnetic field is done by using the principle of "minimal coupling." In classical mechanics, this principle aims at producing Newton's equation with the Lorentz force by constructing a corresponding Lagrangian.

Exercise 3.1 _Study classical minimal coupling by answering the following questions:_

_____

[1] In the literature, this is frequently called semiclassical laser matter interaction [1]. We have, however, used the expression "semiclassics" already differently in Chap. 2.

(a) *Under which conditions does the classical Lagrangian*

$$L = \frac{m}{2}\dot{r}^2 - q\Phi(r,t) + q\dot{r} \cdot A(r,t)$$

lead to Newton's equations of motion with the Lorentz force?
(b) *Give explicit expressions for the canonical momentum* $p = \frac{\partial L}{\partial \dot{r}}$ *and the mechanical momentum* $p_m = m\dot{r}$.
(c) *What is the explicit form of the Hamiltonian* $H = \dot{r} \cdot p - L$?

To arrive at the quantum version of minimal coupling, we could just use the classical result and invoke the correspondence principle. More instructive is a direct approach to quantum minimal coupling, however, which shall be discussed in some detail now. Let us first consider the effect of a local unitary transformation with the scalar field $\chi(r,t)$

$$\Psi'(r,t) = e^{i\frac{q}{\hbar}\chi(r,t)}\Psi(r,t) \tag{3.1}$$

on the time-dependent Schrödinger equation. For the transformed wavefunction, the transformed equation

$$i\hbar\dot{\Psi}'(r,t) = \hat{H}'\Psi'(r,t) \tag{3.2}$$

holds, where the primed Hamiltonian is given by

$$\hat{H}' = e^{i\frac{q}{\hbar}\chi(r,t)}\hat{H}e^{-i\frac{q}{\hbar}\chi(r,t)} - q\dot{\chi} \tag{3.3}$$

with

$$\hat{H} = \frac{\hat{p}^2}{2m} + V(r). \tag{3.4}$$

Shifting the momentum operator $\hat{p} = \frac{\hbar}{i}\nabla$ twice past the exponential factor of the unitary transformation, we get the identity

$$e^{i\frac{q}{\hbar}\chi(r,t)}\hat{p}^2 e^{-i\frac{q}{\hbar}\chi(r,t)}\Psi'(r,t) = (\hat{p} - q\nabla\chi)^2 \Psi'(r,t) \tag{3.5}$$

and therefore

$$\hat{H}' = \frac{1}{2m}\left(\frac{\hbar}{i}\nabla - q\nabla\chi\right)^2 + V(r) - q\frac{\partial\chi}{\partial t} \tag{3.6}$$

holds for the primed Hamiltonian.

Following Weyl [2], the time-dependent Schrödinger equation has to be invariant under the unitary transformation (3.1) introduced above. To satisfy this requirement, the original time-dependent Schrödinger equation has to be modified slightly, however, according to

$$i\hbar\dot{\Psi}(r,t) = \left[\frac{1}{2m}\left(\frac{\hbar}{i}\nabla - qA\right)^2 + V(r) + q\Phi\right]\Psi(r,t). \tag{3.7}$$

This equation is now formally equivalent to the transformed time-dependent Schrödinger equation

$$i\hbar\dot{\Psi}'(r,t) = \left[\frac{1}{2m}\left(\frac{\hbar}{i}\nabla - qA'\right)^2 + V(r) + q\Phi'\right]\Psi'(r,t) \qquad (3.8)$$

if the relations

$$A' = A + \nabla\chi, \qquad \Phi' = \Phi - \dot{\chi} \qquad (3.9)$$

hold. These, however, are the well-known gauge transformations of classical electrodynamics. The electromagnetic fields

$$\mathcal{E} = -\frac{\partial A}{\partial t} - \nabla\Phi \qquad (3.10)$$

$$B = \nabla \times A \qquad (3.11)$$

are unchanged by such a transformation.

Summarizing, minimal coupling amounts to replacing the canonical momentum $\hat{p}$ by the mechanical momentum $\hat{p} - qA(r,t)$ and shifting the potential by $q\Phi$. It is worthwhile to note that the probability flux density in the equation of continuity (2.5) of Chap. 2 has to be changed accordingly.

Exercise 3.2 *Find the modified expression for the probability flux density j in the case of coupling of the motion of a charged particle to an external field. Show that the expression you gained is gauge invariant.*

Expanding the square of the mechanical momentum, cross terms of the form $\hat{p} \cdot A$ and $A \cdot \hat{p}$ appear. In the Coulomb gauge which is defined by

$$\Phi(r,t) = 0, \qquad \nabla \cdot A(r,t) = 0, \qquad (3.12)$$

we can conclude that $\hat{p}$ commutes with A. Therefore, the two cross terms are identical as it is the case in classical mechanics.

In general, the case of a system of many charged particles which are coupled to a laser field has to be studied. As we will see in Chap. 5, the motion of the center of mass and the relative motion without a laser can be separated. With the laser, they do not necessarily separate any more [2]. We will deal with the coupling to a many particle system in more detail in Chap. 5.

3.1.2 Length Gauge

Another well-known form of light matter interaction rests on the dipole approximation, in which case the vector potential is assumed to be independent of position. For an atom (of typical size of the order of Angstroms) in a field of optical wavelength of several hundred nanometers this is a well-founded approximation, as depicted in Fig. 3.1.

Fig. 3.1. An atom in the field of a light wave with wavelength much longer than the typical extension of the atom

The effect of the vector potential in the minimal coupling Hamiltonian of (3.7) in dipole approximation is a time-dependent shift of the momentum in the Hamiltonian

$$\hat{H} = \frac{[\hat{p} - qA(t)]^2}{2m} + V(r) \qquad (3.13)$$

and the term velocity gauge is therefore frequently used. Applying a gauge transformation with the scalar field

$$\chi(r, t) = -r \cdot A(t) \qquad (3.14)$$

leads to transformed potentials of the form

$$A' = 0, \qquad \Phi' = -\frac{\partial \chi}{\partial t} = r \cdot \dot{A} = -r \cdot \mathcal{E}(t), \qquad (3.15)$$

where the last equality follows in Coulomb gauge. The present gauge is thus also called length gauge (electric field couples to the position). The corresponding time-dependent Schrödinger equation then reads

$$i\hbar\dot{\Psi}_{\text{GM}}(r, t) = \left[\frac{\hat{p}^2}{2m} + V(r) - qr \cdot \mathcal{E}(t)\right] \Psi_{\text{GM}}(r, t). \qquad (3.16)$$

Historically, it has been introduced by Göppert-Mayer [3] by using the fact that the Lagrangians in the length as well as in the velocity gauge only differ by a total time-derivative.

Exercise 3.3 *Switch from the velocity to the length gauge by adding a total time derivative to the Lagrangian.*

a) *Show first that adding a total time-derivative $\frac{\mathrm{d}}{\mathrm{d}t}f(r, t)$ to the Lagrangian does not alter the equations of motion.*

b) *In the dipole approximation and the Coulomb gauge ($\Phi = 0, A = A(t)$) add $-q\frac{\mathrm{d}}{\mathrm{d}t}(r \cdot A)$ to the Lagrangian and simplify the resulting expression.*

Velocity and length gauge are equivalent up to a unitary transformation and therefore any measurable quantity may not depend on the gauge used. If one uses approximations along the way, however, there may be orders of magnitude difference between the results predicted in the different gauges. A recent investigation on a gauge-independent strong field approximation is given in [4]. Furthermore, it is worthwhile to note that, as shown in Appendix 3.A, the notion of parity, well known in autonomous Hamiltonian systems, can be generalized to the case of periodically driven systems.

3.1.3 Kramers–Henneberger Transformation

In the case of strong fields, another unitary transformation will turn out to be very useful. We start again from the minimally coupled time-dependent Schrödinger equation (3.7), where we expand the square under the assumption of the Coulomb gauge to arrive at

$$i\hbar\dot{\Psi}(\boldsymbol{r},t) = \left[\frac{1}{2m}\left(\frac{\hbar}{i}\boldsymbol{\nabla} - q\boldsymbol{A}\right)^2 + V(\boldsymbol{r})\right]\Psi(\boldsymbol{r},t)$$

$$= \left[-\frac{\hbar^2}{2m}\boldsymbol{\nabla}^2 + \frac{iq\hbar}{m}\boldsymbol{A}\cdot\boldsymbol{\nabla} + \frac{q^2}{2m}\boldsymbol{A}^2 + V(\boldsymbol{r})\right]\Psi(\boldsymbol{r},t). \quad (3.17)$$

Successively performing two unitary transformations

$$\Psi_{\mathrm{KH}}(\boldsymbol{r},t) = \hat{U}_2\hat{U}_1\Psi(\boldsymbol{r},t) \quad (3.18)$$

with

$$\hat{U}_1 = \exp\left\{\frac{iq^2}{2m\hbar}\int_{-\infty}^t \mathrm{d}t'\,\boldsymbol{A}^2\right\} \quad (3.19)$$

$$\hat{U}_2 = \exp\left\{-\frac{q}{m}\int_{-\infty}^t \mathrm{d}t'\,\boldsymbol{A}\cdot\boldsymbol{\nabla}\right\} \quad (3.20)$$

defines a wavefunction in the Kramers–Henneberger gauge [5, 6]. The first transformation eliminates the squared vector potential, whereas the second one moves the coupling into the argument of the potential, as can be seen by working through Exercise 3.4. The time-dependent Schrödinger equation in the Kramers–Henneberger frame is then given by

$$i\hbar\dot{\Psi}_{\mathrm{KH}}(\boldsymbol{r},t) = \left[-\frac{\hbar^2}{2m}\boldsymbol{\nabla}^2 + V[\boldsymbol{r}+\boldsymbol{\alpha}(t)]\right]\Psi_{\mathrm{KH}}(\boldsymbol{r},t), \quad (3.21)$$

where

$$\boldsymbol{\alpha}(t) = -\frac{q}{m}\int_{-\infty}^t \mathrm{d}t'\,\boldsymbol{A}(t'). \quad (3.22)$$

Exercise 3.4 *Show that the two unitary transformations into the Kramers–Henneberger frame eliminate the terms proportional to $\boldsymbol{A}^2$ and $\boldsymbol{A}$ in the Hamiltonian. Due to the fact that the first transformation is a global phase transformation it just remains to calculate*

$$\hat{U}_2\hat{V}\hat{U}_2^{-1}.$$

Hint: Use the operator relation $e^{\hat{L}}\hat{M}e^{-\hat{L}} = \sum_{n=0}^{\infty}\frac{1}{n!}[\hat{L},\hat{M}]_n$, where $[\hat{L},\hat{M}]_n = [\hat{L},[\hat{L},\hat{M}]_{n-1}]$ and $[\hat{L},\hat{M}]_0 = \hat{M}$.

Differentiating (3.22) twice and using (3.10) in the Coulomb gauge, we find that

$$m\ddot{\alpha}(t) = q\mathcal{E} \qquad (3.23)$$

holds. The Kramers–Henneberger transformation is therefore characterized by a spatial translation into an accelerated frame, corresponding to the oscillatory motion of the charged particle in the electric field. The present gauge is therefore also frequently referred to as the acceleration gauge.

3.1.4 Volkov Wavepacket

In the case of 1d-free motion $V(x) = 0$ of an electron with mass m_e and charge $-e$ in a monochromatic laser field $\mathcal{E} = \mathcal{E}_0 \cos(\omega t)$ the time-dependent Schrödinger equation in length gauge (3.16) can be solved exactly analytically under the assumption of a Gaussian initial state. Due to the fact that the total time-dependent potential is linear, the resulting Volkov wavefunction is given by using the GWD of Sect. 2.1.4 for $\alpha_0 = \gamma/2$ according to[2]

$$\Psi(x,t) = \left(\frac{\gamma}{\pi}\right)^{1/4} \sqrt{\frac{1}{1 + i\gamma\hbar t}} \exp\left\{\frac{i}{\hbar}\left[\frac{U_p}{2\omega}\sin(2\omega t) - U_p t + xp(t)\right]\right\}$$
$$\exp\left\{-\frac{\gamma}{2(1 + i\gamma\hbar t)}(x - q(t))^2\right\}, \qquad (3.24)$$

where the abbreviations

$$p(t) = -e\mathcal{E}_0 \sin(\omega t)/\omega \qquad (3.25)$$
$$q(t) = q_0 + e\mathcal{E}_0[\cos(\omega t) - 1]/(m_e\omega^2) \qquad (3.26)$$

for the solutions of the classical equations of motion for position and momentum have been used and U_p will be defined below. The amplitude of oscillations of position $e\mathcal{E}_0/(m_e\omega^2)$ is the so-called "quiver amplitude". We can convince ourselves of the analytic form of the Volkov solution by working through Exercise 3.5. It will turn out that the change of kinetic energy averaged over a period of the external field vanishes. A free particle can therefore not accumulate energy from the field.

Exercise 3.5 *Using Gaussian wavepacket dynamics calculate the wavefunction of a free particle in a laser field leading to a potential*

$$V_L(x,t) = e\mathcal{E}_0 x \cos(\omega t).$$

(a) *Find the solutions (q_t, p_t) of the classical equations of motion with the initial conditions $(q_0, 0)$. Calculate the kinetic energy and its derivative and average the results over one period $T = 2\pi/\omega$ of the external field. Interpret the results.*

---

[2] The gauge index will be suppressed in the remainder of the book, as we will explicitly state which gauge is used.

(b) Use the result for α_t from the free particle case (why is this possible?).
(c) Using partial integration show that

$$\int_0^t dt' L = -\int_0^t dt' \frac{p_{t'}^2}{2} + q_t p_t$$

holds. Use this result to determine the phase $\delta_t = \int_0^t dt'(L - \alpha_{t'})$ and insert everything in the GWD expression. Why is the final result exact?

As could be seen by working through the previous exercise, there is an important quantity appearing in the Volkov solution. This is the average of the kinetic energy over one period, which is given by

$$U_p = \frac{1}{T} \int_0^T dt \frac{p^2}{2m_e} = \frac{e^2 \mathcal{E}_0^2}{4m_e \omega^2}, \tag{3.27}$$

as can be shown by using (3.25). This quantity is called ponderomotive potential. $2U_p$ is the maximal kinetic energy, that an electron may have at a certain time. For the following, it is important to keep in mind that low frequency fields lead to high ponderomotive potentials.

3.2 Analytically Solvable Two-Level Problems

Driven, two-level systems are the easiest realizations of the field-matter coupling formalism just reviewed. Several paradigms in the theory of laser-induced dynamics are found already in the solutions of these simple systems. They shall therefore now be studied in some detail. We will concentrate on analytically solvable cases which can either be solved exactly or under some approximations.

3.2.1 Dipole Matrix Element

First of all the Hamilton matrix has to be set up. To this end, let us consider two energy levels with the unperturbed orthogonal states $|\psi_1\rangle$ and $|\psi_2\rangle$ with the corresponding energies $E_1 = -\hbar\epsilon, E_2 = \hbar\epsilon$, which are the diagonal elements of the Hamilton matrix.

To write down an expression for the off-diagonal elements of the Hamilton matrix in the case of an external perturbation, we assume that it is due to the coupling to an electric field of the form

$$\mathcal{E}(r,t) = \mathcal{E}_0 \cos(k \cdot r - \omega t). \tag{3.28}$$

We now turn to the dipole approximation of Sect. 3.1.2, i.e., $\lambda = 2\pi/k$ shall be much smaller than the size of the quantum system, as depicted e.g., in Fig. 3.1. In the argument of the cosine r can then be replaced by r_0 which can be set to zero without the loss of generality. The electric field is then purely time-dependent

$$\mathcal{E}(t) = \mathcal{E}_0 \cos(\omega t) \tag{3.29}$$

and the coordinate-independent force

$$F(t) = -e\mathcal{E}(t) \tag{3.30}$$

acts on the electron. The corresponding potential energy is given by

$$\hat{V}_{\mathrm{L}}(r,t) = e\hat{r} \cdot \mathcal{E}(t). \tag{3.31}$$

Adding this potential energy to the Hamiltonian leads to the length gauge form of the Hamiltonian in (3.16).

If the two levels under consideration have eigenfunctions with different parity then

$$\hbar\nu_{12}(t) \equiv \mathcal{E}(t) \cdot \int d^3r\psi_1^*er\psi_2 = \mu_{12} \cdot \mathcal{E}(t) = \hbar\nu_{21}(t) \tag{3.32}$$

follows for the non-vanishing off-diagonal elements of the Hamilton matrix, which are proportional to the matrix element

$$\mu_{12} \equiv \int d^3r\psi_1^*er\psi_2 \tag{3.33}$$

of the dipole operator.

3.2.2 Rabi Oscillations Induced by a Constant Perturbation

For the following, we assume that the perturbation is time-independent, i.e., we set $\omega \to 0$ and define $\nu := \lim_{\omega \to 0} \nu_{12}(t)$. As an Ansatz for the solution of the time-dependent Schrödinger equation in that case, a superposition of the unperturbed eigenstates with time-dependent coefficients

$$|\Psi(t)\rangle = c_1(t)|\psi_1\rangle + c_2(t)|\psi_2\rangle \tag{3.34}$$

can be chosen. For the vector of coefficients, the linear system of coupled ordinary differential equations

$$i\hbar\dot{c} = \mathbf{H}c \tag{3.35}$$

with the two by two Hamilton matrix

$$\mathbf{H} = \hbar \begin{pmatrix} -\epsilon & \nu \\ \nu & \epsilon \end{pmatrix} \tag{3.36}$$

emerges. The eigenvalues of the Hamiltonian are

$$E_\pm = \pm\hbar\sqrt{\epsilon^2 + \nu^2} \tag{3.37}$$

and the corresponding eigenstates are given by

$$|\psi_+\rangle = \sin(\Theta)|\psi_1\rangle + \cos(\Theta)|\psi_2\rangle \tag{3.38}$$
$$|\psi_-\rangle = \cos(\Theta)|\psi_1\rangle - \sin(\Theta)|\psi_2\rangle, \tag{3.39}$$

where the definition

$$\Theta \equiv \frac{1}{2}\arctan(\nu/\epsilon) \tag{3.40}$$

has been used.[3] In the case of degeneracy of the unperturbed states $\epsilon \to 0$, $\Theta = \pi/4$, and the eigenstates are the symmetric, respectively, antisymmetric combination of the two unperturbed states.

The time-evolution operator for the solution of the Schrödinger equation is given by

$$\hat{U}(t,0) = \sum_\pm |\psi_\pm\rangle \exp\left\{-\frac{i}{\hbar}E_\pm t\right\}\langle\psi_\pm|, \tag{3.41}$$

as can be seen by comparison with (2.30). In the basis of the eigenvectors (1,0) and (0,1) of the unperturbed Hamilton matrix, the matrix

$$\mathbf{U}(t,0) = \begin{pmatrix} \sin^2(\Theta) & \sin(\Theta)\cos(\Theta) \\ \sin(\Theta)\cos(\Theta) & \cos^2(\Theta) \end{pmatrix}\exp\left\{-\frac{i}{\hbar}E_+ t\right\}$$
$$+ \begin{pmatrix} \cos^2(\Theta) & -\sin(\Theta)\cos(\Theta) \\ -\sin(\Theta)\cos(\Theta) & \sin^2(\Theta) \end{pmatrix}\exp\left\{-\frac{i}{\hbar}E_- t\right\}$$

for the time-evolution operator can be derived. This matrix allows us to calculate

$$P_{21}(t) = |\langle\psi_2|\hat{U}(t,0)|\psi_1\rangle|^2 = |U_{21}(t,0)|^2, \tag{3.42}$$

which is the probability to find the system in state $|\psi_2\rangle$ at time t, if it was in state $|\psi_1\rangle$ at time zero (in terms of the coefficients this corresponds to the initial conditions $c_1(0) = 1, c_2(0) = 0$). From the matrix representation of $\hat{U}$

$$P_{21}(t) = \frac{\nu^2}{\nu^2 + \epsilon^2}\sin^2(\Omega_R t/2) \tag{3.43}$$

follows, where

$$\Omega_R \equiv 2\sqrt{\epsilon^2 + \nu^2} \tag{3.44}$$

---

[3] $\arctan(x) = \arccos(1/\sqrt{1+x^2})$ and $\arctan(x) = \arcsin(x/\sqrt{1+x^2})$ can be used to resolve the cosine and sine terms in (3.38,3.39).

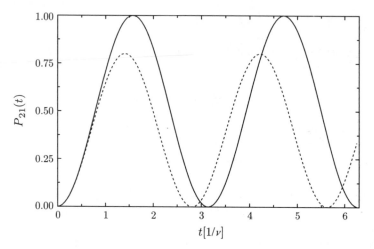

Fig. 3.2. Rabi oscillations of the probability to be in the upper state, starting from the lower state, induced by the perturbation $\nu = 1$. The degenerate $\epsilon = 0$ (*solid line*), as well as the non-degenerate case $\epsilon = 0.5$ (*dashed line*) are depicted as a function of time in units of $1/\nu$; all energies in arbitrary units

is the definition of the so-called Rabi frequency. $P_{21}(t)$ performs Rabi oscillations with the amplitude $A = \frac{\nu^2}{\nu^2+\epsilon^2}$, which are depicted in Fig. 3.2. Only in case of degeneracy, $\epsilon = 0$, do the oscillations have an amplitude of 1. Furthermore, for non-degenerate systems the oscillations are faster than for degenerate unperturbed levels.

Rabi oscillations are analogous to the tunneling dynamics of the probability amplitude in a symmetric double well, which will be considered in Chap. 5. There the eigenstates of the unperturbed problem are the symmetric, respectively, antisymmetric superposition of *localized* states in the left and right well and thus for a localized initial condition $c_1 = \pm c_2 = 1/\sqrt{2}$ has to be chosen.

3.2.3 Time-Dependent Perturbations

In the presence of a time-dependent perturbation $\hat{V}(t) = \hbar\hat{v}(t)$ the time-dependent Schrödinger equation for the coefficients is

$$i\dot{c}_1 = c_1\epsilon_1 + c_2\nu_{12}(t) \tag{3.45}$$
$$i\dot{c}_2 = c_2\epsilon_2 + c_1\nu_{21}(t) \tag{3.46}$$

In the "strong-coupling"-limit, i.e., for $\nu_{21} \gg \epsilon_2 - \epsilon_1$ these coupled differential equations can be solved perturbatively [7]. However, there is an another, approximate approach to solve the differential equations, starting from the Ansatz

$$c_1(t) = d_1(t)\exp[-i\epsilon_1 t] \tag{3.47}$$
$$c_2(t) = d_2(t)\exp[-i\epsilon_2 t], \tag{3.48}$$

which leads to

$$i\dot{d}_1 = d_2 \nu_{12}(t) \exp[-i\omega_{21}t], \tag{3.49}$$

$$i\dot{d}_2 = d_1 \nu_{21}(t) \exp[i\omega_{21}t], \tag{3.50}$$

where the abbreviation $\omega_{21} = \epsilon_2 - \epsilon_1$ has been introduced. Note that the transformation from the vector c to the vector d of coefficients is equivalent to a unitary transformation into the interaction picture.

In the case of a monochromatic (coherent) perturbation the system of differential equations can be solved analytically by using the so-called rotating wave approximation, as will be shown in the following. In the case of interaction with incoherent radiation (a random superposition of monochromatic laser fields) we can use perturbation theory and in this way give a microscopic derivation of the B-coefficient of Chap. 1. This last case will be dealt with in Appendix 3.B.

Rotating Wave Approximation

For the monochromatic field in (3.29), the Schrödinger equation in the interaction picture (3.49), (3.50) can be written as

$$i\dot{d}_1 = d_2 \frac{\mu_{12} \cdot \mathcal{E}_0}{2\hbar} (\exp[i(\omega - \omega_{21})t] + \exp[-i(\omega + \omega_{21})t]) \tag{3.51}$$

$$i\dot{d}_2 = d_1 \frac{\mu_{21} \cdot \mathcal{E}_0}{2\hbar} (\exp[-i(\omega - \omega_{21})t] + \exp[i(\omega + \omega_{21})t]). \tag{3.52}$$

To introduce the rotating wave approximation (RWA), we define the detuning between the field and the external frequency

$$\Delta_d \equiv \omega - \omega_{21}. \tag{3.53}$$

For $\Delta_d \ll \omega_{21}$, the terms that oscillate at about twice the frequency of the external field are the so-called counter-rotating terms. In the differential equations given above they can be neglected if we assume that the coefficients $d_{1,2}$ change on a much longer time scale. To prove this procedure mathematically, one has to average the differential equations over times much larger than $1/(\omega + \omega_{21})$.

The differential equations in RWA are now dramatically simplified and read

$$i\dot{d}_1 = d_2 \frac{\mu\mathcal{E}_0}{2\hbar} \exp[i\Delta_d t] \tag{3.54}$$

$$i\dot{d}_2 = d_1 \frac{\mu\mathcal{E}_0}{2\hbar} \exp[-i\Delta_d t], \tag{3.55}$$

where we have assumed in addition, that the polarization of the field is in the direction of the dipole matrix element, which has the absolute value $\mu = \mu_{12} = \mu_{21}$. The solution of the two coupled differential equations can be found by differentiating (3.54) and inserting it into (3.55). The second-order differential equation that emerges can be solved and one gets

$$d_1(t) = \frac{\hbar}{\mu\mathcal{E}_0} \exp[i\Delta_{\mathrm{d}} t/2]\{(\Delta_{\mathrm{d}} - \Omega_{\mathrm{R}})C \exp[i\Omega_{\mathrm{R}} t/2]$$
$$+ (\Delta_{\mathrm{d}} + \Omega_{\mathrm{R}})D \exp[-i\Omega_{\mathrm{R}} t/2]\} \tag{3.56}$$

$$d_2(t) = \exp[-i\Delta_{\mathrm{d}} t/2]\{C \exp[i\Omega_{\mathrm{R}} t/2] + D \exp[-i\Omega_{\mathrm{R}} t/2]\}, \tag{3.57}$$

with the Rabi frequency

$$\Omega_{\mathrm{R}} = \sqrt{\Delta_{\mathrm{d}}^2 + \left(\frac{\mu\mathcal{E}_0}{\hbar}\right)^2}. \tag{3.58}$$

The parameters C and D have to be determined from the initial conditions. In the case of non-resonance (corresponding to the non-degenerate case for constant perturbations), the oscillations are again faster than on resonance.

Exercise 3.6 *Consider a two-level system interacting with a monochromatic laser field.*

(a) Average the TDSE over times long in comparison to $1/(\omega + \omega_{21})$ to motivate neglecting the counter-rotating terms.

(b) Using the initial conditions $d_1(0) = 1$ and $d_2(0) = 0$ give explicit expressions for C and D. Depict $|d_2(t)|^2$ for resonance as well as off resonance.

Furthermore, the quality of the RWA depends on the quality of the argument of neglecting the counter-rotating terms. The validity of this assumption can be studied explicitly for a specific example in Exercise 3.7.

Exercise 3.7 *An electron shall move in an inversion symmetric potential $V(x) = V(-x)$ in one spatial dimension.*

(a) Show that the eigenfunctions of the TISE fulfill

$$\psi_{2n}(x) = \psi_{2n}(-x) \qquad \text{resp.} \qquad \psi_{2n+1}(x) = -\psi_{2n+1}(-x)$$

and that diagonal dipole matrix elements $\mu_{nn} = \langle\psi_n|e\hat{x}|\psi_n\rangle$ therefore always vanish.

(b) Calculate the dipole matrix element between the ground and the first excited state of the harmonic oscillator

$$V(x) = \frac{1}{2}m_{\mathrm{e}}\omega_{\mathrm{e}}^2 x^2$$

with a frequency in the visible range around $\omega_{\mathrm{e}} = 3.14\times10^{15}\mathrm{s}^{-1}$. Determine the Rabi frequency in the resonance case for three different field strengths $\mathcal{E}_0 = 1, 10^6, 10^{10}\mathrm{Vcm}^{-1}$. Is the condition for the applicability of the RWA fulfilled for all field strengths?

Finally, let us consider the case of resonance, $\Delta_d = 0$, in which the external frequency equals the level spacing. Furthermore, we assume that the external field shall be of finite duration, i.e., it shall consist of a laser pulse with an envelope, so that we have to replace $\mathcal{E}_0$ by $\mathcal{E}_0 f(t)$ in the time-dependent Schrödinger equation. According to (3.58), a time-dependent Rabi frequency

$$\Omega_R(t) = \frac{\mu \mathcal{E}_0 f(t)}{\hbar} \tag{3.59}$$

emerges, with the help of which the coupled differential equations can be written as

$$i\dot{d}_1 = \frac{\Omega_R(t)}{2} d_2, \tag{3.60}$$

$$i\dot{d}_2 = \frac{\Omega_R(t)}{2} d_1. \tag{3.61}$$

For the initial conditions $d_1(0) = 1, d_2(0) = 0$ the solutions are given by

$$d_1(t) = \cos\left(\int_0^t dt' \frac{\Omega_R(t')}{2}\right) \tag{3.62}$$

$$d_2(t) = -i \sin\left(\int_0^t dt' \frac{\Omega_R(t')}{2}\right), \tag{3.63}$$

as can be verified by insertion. In RWA the population transfer in the resonance case does not depend on the specific shape of the pulse, but only on the area below the pulse. This is the so-called area theorem. A π-pulse e.g., allows for a complete transfer of population.

3.2.4 Exactly Solvable Time-Dependent Cases

In very few special cases, also in the case of a time-dependent perturbation an exact analytical solution of the time-dependent two-level Schrödinger equation can be found [8]. As our starting point, we use (3.35) in the case of time-dependent ϵ and ν. After elimination of c_1, the equation

$$\ddot{c}_2 - \frac{\dot{\nu}}{\nu}\dot{c}_2 + \left(\epsilon^2 + \nu^2 - i\dot{\epsilon} + i\epsilon\frac{\dot{\nu}}{\nu}\right) c_2 = 0 \tag{3.64}$$

for the other coefficient can be derived, see e.g., Exercise 3.8. Reasons for the time-dependence of the diagonal as well as for the off-diagonal elements of the Hamiltonian may be the coupling to a light field or nuclear motion in a molecule, which will be considered in detail in Chap. 5.

In the case of coupling to a pulsed laser field and in RWA[4] we can e.g., choose

$$\epsilon = \Delta/2, \qquad \nu(t) = \nu_0 \mathrm{sech}(t/T_\mathrm{p}), \tag{3.65}$$

defining the Rosen–Zener model with a pulse length parameter T_p. As found by these authors, the solution of the time-dependent Schrödinger equation for this choice can be determined exactly, analytically. With the initial condition $c_1(-\infty) = 1$ and for $t \to \infty$ it is given by [9]

$$|c_2(\infty)|^2 = \sin^2(\pi\nu_0 T_\mathrm{p})\mathrm{sech}^2(\pi\Delta T_\mathrm{p}/2). \tag{3.66}$$

For the resonance case, $\Delta = 0$ you can convince yourself of this solution by working through Exercise 3.8. In the resonance case, it is also rewarding to note that in the argument of the sine, the pulse area $\nu_0 \int_{-\infty}^{\infty} dt\mathrm{sech}(t/T_\mathrm{p}) = \nu_0\pi T_\mathrm{p}$ appears. This is yet another manifestation of the area theorem discussed at the end of Sect. 3.2.3.

Exercise 3.8 *Consider the TDSE for the two-level Rosen–Zener model.*

(a) Prove the equation for c_2 that can be gained by the elimination of c_1.
(b) Transform the independent variable with the help of

$$z = \frac{1}{2}\left(\tanh\frac{t}{T_\mathrm{p}} + 1\right).$$

What is the differential equation for $c_2(z)$?
(c) Consider the special case $\epsilon = 0$ and determine $c_2(t = \infty)$ for the initial conditions $c_2(t = -\infty) = 0$ and $c_1(-\infty) = 1$.
Hint: Use the hypergeometric function (see e.g., I.S. Gradshteyn and I.M. Rhyzik, Tables of Integrals Series and Products (Academic Press, San Diego, 1994), Sect. 9.1) and $F(a, b, c, 1) = \frac{\Gamma(c)\Gamma(c-a-b)}{\Gamma(c-a)\Gamma(c-b)}$, $\Gamma(1-x)\Gamma(x) = \frac{\pi}{\sin(\pi x)}$, $\Gamma(3/2) = \sqrt{\pi}/2$.

Apart from the Rosen–Zener solution an exact analytic solution is available in the case of the time-dependence being induced by moving nuclei, which can be modeled by

$$\epsilon(t) = \lambda t, \qquad \nu(t) = \nu_0. \tag{3.67}$$

This case has been investigated by Landau [10] as well as by Zener [11] and an asymptotic solution of (3.64) for the initial condition $c_1(-\infty) = 1$ is

$$|c_2(\infty)|^2 = 1 - \exp[-2\pi\gamma], \tag{3.68}$$

where $\gamma = \nu^2/|2\dot{\epsilon}|$. The expression for γ can be further specified in molecular theory and leads to the celebrated Landau–Zener formula [12].

---

[4] This is an approximation and therefore the notion of exact solubility refers to the final equation and not the initial problem.

3.A Generalized Parity Transformation

In the case of a symmetric static potential $V(x) = V(-x)$ and in length gauge with a sinusoidal laser potential of the form $e\mathcal{E}_0 x \sin(\omega t)$, the extended Hamiltonian $\mathcal{H}$ in (2.117) is invariant under the generalized parity transformation

$$\mathcal{P}: \qquad x \to -x, \quad t \to t + \frac{T}{2}. \qquad (3.69)$$

The Floquet functions thus transform according to

$$\mathcal{P}\psi_{\alpha'}(x,t) = \pm\psi_{\alpha'}(x,t), \qquad (3.70)$$

i.e., they have either positive or negative generalized parity. With the help of (2.129) it follows that $\psi_{\alpha'}(x,t), \psi_{\beta'}(x,t)$ have the same or different generalized parity, depending on $(\alpha - k) - (\beta - l)$ being odd or even.

As we will see in Chap. 5, exact crossings of the quasienergies as a function of external parameters are of utmost importance for the quantum dynamics of periodically driven systems. For stationary systems, the possibility of exact crossings has been studied in the heyday of quantum theory by von Neumann and Wigner [13]. These authors found that eigenvalues of eigenfunctions with different parity may approach each other arbitrarily closely and may thus cross exactly. This is in contrast to eigenvalues of the same parity, which always have to be at a finite distance, a fact which is sometimes referred to as the non-crossing rule. The corresponding behavior in the spectrum as a function of external parameters is called allowed, respectively, avoided crossing. In the Floquet case, the Hamiltonian can also be represented by a Hermitian matrix, see e.g., (2.158), and therefore the same reasoning applies with parity replaced by generalized parity.

For the investigations to be presented in Sect. 5.4.1 it is decisive if these *exact* crossings are singular events in parameter space or if they can occur by variation of just a single parameter. In [13], it has been shown that for Hermitian matrices (of finite dimension) with complex (real) elements the variation of three (two) free parameters is necessary in order for two eigenvalues to cross. Using similar arguments, it can be shown that for a real Hermitian matrix with alternatingly empty off-diagonals (as it is e.g., the case for the Floquet matrix of the periodically driven, quartic, symmetric, bistable potential) the variation of a single parameter is enough to make two quasienergies cross.

In the case of *avoided* crossings an interesting behavior of the corresponding eigenfunctions can be observed. There is a continuous change in the structure in position space if one goes through the avoided crossing [14]. Pictorially, this is very nicely represented in the example of the driven quantum well, depicted in Fig. 3.3, taken out of [15], where for reasons of better visualization the Husimi transform of the quasi-eigenfunctions as a function of action angle variables (J, Θ) [16] is shown.

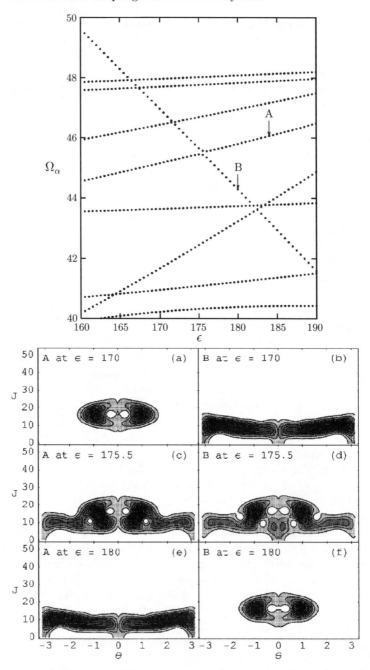

Fig. 3.3. Avoided crossing of Floquet energies (here denoted by Ω_α) as a function of field amplitude (*upper panel*) and associated change of character of the Floquet functions in the driven quantum well (*lower panels* **(a)-(f)**); from [15]

3.B Two-Level System in an Incoherent Field

In the discussion of time-dependently driven two-level systems, it was mentioned that a two-level system in an incoherent external field can be treated in perturbation theory.

As the starting point, we use the Schrödinger equation in the interaction representation (3.49), (3.50) with the initial conditions $d_1(0) = 1$ and $d_2(0) = 0$. For very small perturbations, the coefficient d_1 is assumed to remain at its initial value, leading to

$$i\dot{d}_2 = \nu_{21}(t)\exp[i\omega_{21}t]. \tag{3.71}$$

This equation can be integrated immediately to yield

$$d_2 = -i\int_0^t dt'\nu_{21}(t')\exp[i\omega_{21}t']. \tag{3.72}$$

The field shall consist of a superposition of waves with uniformly distributed, statistically independent phases Φ_j

$$\mathcal{E}(t) = \frac{1}{2}\sum_{\omega_j>0}\mathcal{E}_j\exp[i\phi_j - i\omega_j t] + \text{c.c.}. \tag{3.73}$$

If we insert this into the equation above, we get

$$d_2 = -\frac{i}{2\hbar}\sum_j\mathcal{E}_j\cdot\boldsymbol{\mu}_{21}\exp[i\phi_j]\int_0^t dt'\exp[i(\omega_{21}-\omega_j)t'] \tag{3.74}$$

$$= -\frac{i}{2\hbar}\sum_j\mathcal{E}_j\cdot\boldsymbol{\mu}_{21}\exp[i\phi_j]S_j, \tag{3.75}$$

where the definition

$$S_j = [i(\omega_{21}-\omega_j)]^{-1}\{\exp[i(\omega_{21}-\omega_j)t]-1\} \tag{3.76}$$

has been introduced. The occupation probability of the second level is then given by the double sum

$$|d_2|^2 = (2\hbar)^{-2}\sum_j\sum_{j'}\exp[i(\phi_j-\phi_{j'})]\mathcal{E}_j\cdot\boldsymbol{\mu}_{21}\mathcal{E}_{j'}\cdot\boldsymbol{\mu}_{21}^*S_jS_{j'}^*. \tag{3.77}$$

Averaging over the phases is now performed and denoted by $\langle\,\rangle$, yielding

$$\langle\exp[i(\phi_j-\phi_k)]\rangle = \delta_{jk}. \tag{3.78}$$

One of the sums in (3.77) collapses and

$$\langle|d_2(t)|^2\rangle = \left|\frac{\boldsymbol{e}\cdot\boldsymbol{\mu}_{21}}{\hbar}\right|^2\sum_j|\mathcal{E}_j|^2(\omega_{21}-\omega_j)^{-2}\sin^2[(\omega_{21}-\omega_j)t/2] \tag{3.79}$$

follows for identical polarization of the light waves $\boldsymbol{e}$.

Now, we have to sum over the distribution of frequencies. To this end, we consider the time derivative of the expression above[5]

$$\frac{\mathrm{d}}{\mathrm{d}t}\langle|d_2(t)|^2\rangle = \left|\frac{e\cdot\boldsymbol{\mu}_{21}}{\sqrt{2\hbar}}\right|^2 \sum_j |\mathcal{E}_j|^2 (\omega_{21}-\omega_j)^{-1}\sin[(\omega_{21}-\omega_j)t]. \quad (3.80)$$

With the definition of an energy density per *angular* frequency interval $\rho(\omega_j) = \frac{1}{2}\epsilon_0|\mathcal{E}_j|^2/\Delta\omega_j$, assuming that the frequencies are distributed continuously, and replacing $\rho(\omega_j)$ by its resonance value $\rho(\omega_{21})$, due to

$$\int_{-\infty}^{\infty} \mathrm{d}\omega\, \sin(\omega t)/\omega = \pi, \quad (3.81)$$

we get

$$\frac{\mathrm{d}}{\mathrm{d}t}\langle|d_2(t)|^2\rangle = \frac{\pi}{\epsilon_0}\left|\frac{e\cdot\boldsymbol{\mu}_{21}}{\hbar}\right|^2 \rho(\omega_{21}). \quad (3.82)$$

The right-hand side of this expression is a constant and therefore consistent with the assumptions made in the derivation of Planck's radiation law in Chap. 1.

Comparing the equation above with (1.2) in the case $N_1=1$ and after switching to the *linear* frequency case [17]

$$B_{12} = \frac{2\pi^2}{\epsilon_0}\left|\frac{e\cdot\boldsymbol{\mu}_{21}}{h}\right|^2 \quad (3.83)$$

is found for Einstein's B coefficient.

Notes and Further Reading

The theory of minimal coupling and the different gauges related by unitary transformations are at least partly covered in many textbooks. Schleich's book [2] focuses on the subtleties arising from the inclusion of center of mass motion and contains an appendix dealing with terms beyond the dipole approximation.

Our formulation of the interaction of two-level systems with coherent and incoherent light is based on the presentation in Haken's book [18]. A landmark paper in this field is the one by Shirley [19], treating the periodically driven two-level problem in Floquet theory. The theory of two-level systems interacting with magnetic fields has not been dealt with here but is covered in the book by Weissbluth [20]. This book is also a treasure-house if one is interested in the effect of damping on the dynamics of a two-level system. The wide field of dissipation in quantum theory requires a treatment in terms of the density matrix. More details on that exciting field can be found in the book by Weiss [21].

---

[5] Note that $2\cos(x/2)\sin(x/2) = \sin(x)$.

The laser field is considered to be a classical field throughout the rest of this book. In quantum optics, where the light field is treated quantum mechanically, the RWA can also be performed and if applied to a driven two-level system this is known as the Jaynes–Cummings model, which is treated in detail e.g., in the book by Schleich [2].

References

1. A.D. Bandrauk, in *Molecules in Laser Fields*, ed. by A.D. Bandrauk (Dekker, New York, 1994), chap. 1, pp. 1–69
2. W.P. Schleich, *Quantum Optics in Phase Space* (Wiley, Berlin, 2000)
3. M. Göppert-Mayer, Ann. Phys. (Leipzig) **9**, 273 (1931)
4. F.H.M. Faisal, Phys. Rev. A **75**, 063412 (2007)
5. H.A. Kramers, *Collected Scientific Papers* (North Holland, Amsterdam, 1956)
6. W.C. Henneberger, Phys. Rev. Lett. **21**, 838 (1968)
7. J.C.A. Barata, W.F. Wreszinski, Phys. Rev. Lett. **84**, 2112 (2000)
8. S. Stenholm, in *Quantum Dynamics of Simple Systems*, ed. by G.L. Oppo, S.M. Barnett, E. Riis, M. Wilkinson (IOP, Bristol, 1996), p. 267
9. N. Rosen, C. Zener, Phys. Rev. **40**, 502 (1932)
10. L.D. Landau, Phys. Z. Soviet Union **2**, 46 (1932)
11. C. Zener, Proc. Roy. Soc. (London) A **137**, 696 (1932)
12. D. Coker, in *Computer Simulation in Chemical Physics*, ed. by M.P. Allen, D.J. Tildesley (Kluwer, Amsterdam, 1993)
13. J. von Neuman, E. Wigner, Phys. Z. **30**, 467 (1929)
14. A.G. Fainshteyn, N.L. Manakov, L.P. Rapoport, J. Phys. B **11**, 2561 (1978)
15. T. Timberlake, L.E. Reichl, Phys. Rev. A **59**, 2886 (1999)
16. H. Goldstein, *Klassische Mechanik* (Aula-Verlag, Wiesbaden, 1985)
17. R.C. Hilborn, Am. J. Phys. **50**, 982 (1982)
18. H. Haken, *Licht und Materie Bd. 1: Elemente der Quantenoptik* (BI Wissenschaftsverlag, Mannheim, 1989)
19. J.H. Shirley, Phys. Rev. **138**, B979 (1965)
20. M. Weissbluth, *Photon–Atom Interactions* (Academic, New York, 1989)
21. U. Weiss, *Quantum Dissipative Systems*, 2nd edn. (World Scientific, Singapore, 1999)

4

Single Electron Atoms in Strong Laser Fields

The interaction of light fields with atoms has a long history. A classic reference is the book "Photon–Atom Interactions" by Weissbluth [1]. The focus in that book is on the interaction of atoms with weak fields, leading to absorption or emission of only few quanta of radiation. In this chapter, some modern applications of laser–matter interaction in the field of atomic physics shall be studied, however. Due to the availability of short and strong laser pulses a range of new and partly counterintuitive phenomena can be observed. Some of these are:

- Above threshold ionization (ATI)
- Multiphoton ionization (MPI)
- Localization of Rydberg atoms by half cycle pulses (HCP)
- High harmonic generation (HHG)

These phenomena can in general not be understood in the framework of perturbation theory. It turns out that the time-dependent wavepacket approach of Chap. 2 is the method of choice to describe and understand a lot of new atomic physics in strong laser fields. In this chapter, we will almost exclusively deal with the dynamics of a *single* electron initially bound in a Coulomb potential. The discussion will therefore begin with a short review of the unperturbed hydrogen atom.

4.1 The Hydrogen Atom

In the beginning of this chapter, some well-known results from basic quantum mechanics courses are gathered. These are the eigenvalues and eigenfunctions of the 3d hydrogen atom. Many numerical studies are performed in 1d and therefore also some less familiar eigensolutions in one dimension will be reproduced here.

4.1.1 Hydrogen in Three Dimensions

The simplest atomic problem is that of the hydrogen atom, where a single electron feels the bare Coulomb potential

$$V_C \equiv -\frac{1}{4\pi\varepsilon_0}\frac{e^2}{r} \tag{4.1}$$

of a proton. This problem is solvable exactly analytically and therefore is at the heart of every basic quantum mechanics course. We assume familiarity of the reader with the solution procedure and only reproduce the final results here.

Eigenvalues and Eigenfunctions

Under the approximation of an infinite nuclear mass (i.e., for $M_p/m_e \approx \infty$[1]) the eigenvalues and eigenfunctions for $E < 0$ are given by [2]

$$E_n = -\frac{e^2}{8\pi\varepsilon_0 a_0}\frac{1}{n^2}, \qquad n = 1, 2, \ldots, \tag{4.2}$$

$$\psi_{n,l,m}(r,\Theta,\phi) = R_{n,l}(r)Y_{lm}(\Theta,\phi), \tag{4.3}$$

where the Bohr radius

$$a_0 = 4\pi\varepsilon_0\frac{\hbar^2}{m_e e^2}. \tag{4.4}$$

has been introduced, spherical coordinates (r,Θ,ϕ) have been used, and Y_{lm} are the spherical harmonics. The radial function is of the form

$$R_{n,l}(r) = N_{n,l}\exp(-r'/n)\left(\frac{2r'}{n}\right)^l L_{n-l-1}^{2l+1}\left(\frac{2r'}{n}\right) \tag{4.5}$$

with the (associated) Laguerre polynomials[2]

$$L_k^s(x) = \sum_{\nu=0}^{k}\frac{(k+s)!}{(k-\nu)!(s+\nu)!}\frac{(-x)^\nu}{\nu!}, \tag{4.6}$$

the dimensionless radius $r' = r/a_0$, and a normalization factor

$$N_{nl} = \frac{2}{n^2}\sqrt{\frac{(n-l-1)!}{(n+l)!}}a_0^{-3/2}. \tag{4.7}$$

---

[1] Taking the finiteness of the proton mass into account would lead to the replacement of the electron mass by the reduced mass $\mu = m_e M_p/(m_e + M_p)$ in the final solution.

[2] Please note that we are using the definition of [3] which leads to a slightly different normalization factor as compared to the one in [2].

The radial quantum number $n_r = n - l - 1$ gives the number of nodes (apart from the origin) of the radial function, while the angular momentum quantum number fulfills $0 \leq l \leq n - 1$ and the magnetic quantum number m runs between $-l$ and $+l$. For future reference we note that the Laguerre polynomials are related to the confluent hypergeometric functions according to [3]

$$L_k^s(x) = \binom{k+s}{k} {}_1F_1(-k; s+1; x). \qquad (4.8)$$

For the nodeless ground state wavefunction of hydrogen, which is also referred to as the 1s wavefunction, we use that $L_0^1 = 1$ and $Y_{00} = 1/\sqrt{4\pi}$ in order to arrive at

$$\psi_{1,0,0}(r, \Theta, \phi) = \frac{1}{\sqrt{\pi a_0^3}} \exp\{-r'\}, \qquad (4.9)$$

which will be needed again in Chap. 5.

Atomic Units

To simplify the notation considerably, from now on we will almost exclusively use atomic units (a.u.). They are defined by

$$\hbar = e = m_e = a_0 = 1 \text{ a.u.} \qquad (4.10)$$

Using combinations of powers of these units one can construct atomic units for other observables, as can be seen by looking at Table 4.1 in Appendix 4.A, where SI and atomic units are given for some frequently occurring physical quantities.

From the definition (4.4), the Bohr radius in SI units follows to be $a_0 \approx 0.53 \times 10^{-10}$ m. Other SI base units as those of time and current are given by $1\,\mathrm{s} \approx 4.13 \times 10^{16}$ a.u. and $1\,\mathrm{A} \approx 151$ a.u. The atomic unit of time therefore is of the order of some tens of attoseconds. The relation of atomic units to expectation values of the 3d hydrogen problem is also elucidated in Appendix 4.A.

4.1.2 The One-Dimensional Coulomb Problem

After reviewing the hydrogen atom in full dimensionality, we now concentrate on its one-dimensional analog which is very frequently used for numerical studies.

Exact 1d Coulomb potential

In atomic units the exact or "bare" 1d Coulomb potential is given by

$$V(x) = -\frac{1}{|x|} \qquad (4.11)$$

and similar to the 3d potential has the problem of being singular at the origin. Fortunately, the eigenvalues of the singular potential can be determined analytically [4] and are given by

$$E_n = -\frac{1}{2n^2} \qquad n = 1, 2, 3 \ldots. \tag{4.12}$$

They are converging to the "continuum threshold" $E = 0$ and are equivalent to the eigenvalues of the 3d problem. In addition, however, in the 1d case there is also an eigenvalue $E_0 = -\infty$, corresponding to a δ-function type eigenfunction. If the initial state to be propagated is chosen such that it is zero at the origin then the unphysical state is eliminated from the dynamics [5].

The eigenfunctions of the nonsingular eigenvalues, which are doubly degenerate, are given by [4]

$$\psi_n^\sigma = \sqrt{2/n^3}|x|(\text{sign}(x))^\sigma \exp\left\{-\frac{|x|}{n}\right\} {}_1F_1\left[1 - n; 2; \frac{2|x|}{n}\right], \tag{4.13}$$

where ${}_1F_1$ is the confluent hypergeometric function and where the similarity to the radial functions in the 3d-case for $l = 0$ is obvious. For a given n there are two eigenfunctions, one with positive and one with negative parity ($\sigma = 0$: even, $\sigma = 1$: odd). This seems to be contradicting the "theorem" that there is no degeneracy in 1d quantum spectra. The theorem, however, can only be derived under the assumption that the potential has no singularities, which is not true for the bare Coulomb problem [4]!

Furthermore, a short remark on the 1d Coulomb potential restricted to the half space $x > 0$ shall be made. It is given by

$$V(x) = \begin{cases} \infty & x \leq 0 \\ -\frac{1}{x} & x > 0 \end{cases} \tag{4.14}$$

and in suitable units describes the problem of a "surface state electron" [6]. The spectrum is again identical to the 3d case. Surface state electrons are hovering above a dielectric surface and are especially interesting in the context of quantum chaos if they are driven by microwave radiation. In [6], this is discussed in great detail.

Soft-Core Coulomb Potential

For ease of computation nonsingular approximations to the bare (hard-core) Coulomb potential are used frequently. These are created by adding a constant term in the numerator, leading to

$$V_{\text{sc}}(x) \equiv -\frac{1}{\sqrt{x^2 + a}}, \tag{4.15}$$

for the dimensionless "soft-core" Coulomb potential, thereby introducing a length scale into the problem already in the potential [7].[3] Frequently the

---

[3] The length scale a_0 in the 3d Coulomb problem appears in the solution.

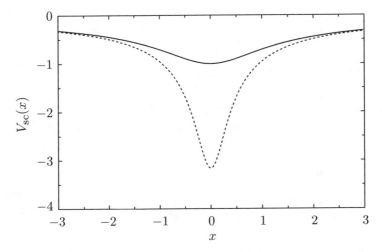

Fig. 4.1. Soft-core Coulomb potential with $a = 1$ (*solid line*) vs. $a = 0.1$ (*dashed line*) as a function of x in atomic units

atomic scale of the Bohr radius is used for a in dimensioned units and therefore $a = 1$ in the dimensionless case. Another potential can be gained by choosing $a = 2$, which leads to a ground state energy of -0.5, equivalent to the bare 3d case. Choosing very small values of the parameter a, the bare Coulomb potential is approached, see Fig. 4.1.

In Sect. 4.2, numerical results for the soft-core as well as for the bare Coulomb potential will be compared.

4.2 Field Induced Ionization

Apart from the excitation of higher lying states, the most prominent effect that an external field can have on the dynamics of a single electron atom is ionization. We will deal with relatively strong fields in the following and will thereby focus on ionization phenomena in several different regimes, ranging from the quasistationary case to the case of almost δ-function like perturbation by HCP fields.

4.2.1 Tunnel Ionization

In the case of long wavelengths and strong fields ($I \approx 10^{14}\,\mathrm{W\,cm^{-2}}$) the ionization of an atom induced by a laser can be treated under the assumption of a quasistationary field. A snapshot of the distorted potential

$$V(\boldsymbol{r}, t) = V_{\mathrm{C}}(\boldsymbol{r}) + \boldsymbol{r} \cdot \boldsymbol{\mathcal{E}}_0 \sin(\omega t) \qquad (4.16)$$

at $t = -\pi/(2\omega)$ with the electric field polarized in the z-direction is shown in Fig. 4.2. Due to the low barrier, the electron can tunnel out of the region

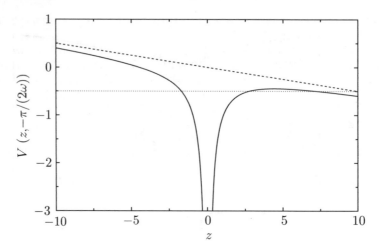

Fig. 4.2. Statically distorted Coulomb potential (*solid line*) along the z-axis and potential induced by an external field of amplitude $\mathcal{E}_0$=0.05 a.u. (*dashed line*). The ground state energy of the unperturbed hydrogen atom in atomic units is indicated by the *horizontal dotted line*

of attraction relatively easily. For stronger fields a direct "over the barrier" ionization would even be possible. The experimental realization of tunnel ionization is usually done with the help of infrared lasers like the CO_2-Laser (with a wave length of 10.6 μm) [8].

In the case of the 3d hydrogen atom, the ionization rate for the case of $\mathcal{E}_0 \ll 1$ is given by (see exercise 1 on page 283 in [9]),

$$\Gamma_{\text{tu}} = \frac{4}{\mathcal{E}_0} \exp\left\{-\frac{2}{3\mathcal{E}_0}\right\}. \tag{4.17}$$

A generalization of this formula for arbitrary atoms has been given by Ammosov, Delone, and Krainov [10]. An easy derivation of an analogous form of the exponential dependence on inverse field strength is possible, if we consider tunneling out of a distorted square potential well of zero range, which for better visibility can also be thought of as a finite range potential.

Exercise 4.1 *Calculate the Gamov factor*

$$\Gamma_{\text{tu}} \sim \exp\left[-2^{3/2} \int_0^{r_0} dr \sqrt{V(r) - E_{\text{g}}}\right]$$

for the rate of tunneling at a negative energy E_{g} out of a bound state of a tilted square potential well (with $V(r=0)=0$) in the presence of a quasistatic external field $\mathcal{E}_0$ (see Fig. 4.3) and compare the result with the exponential factor in the case of 3d hydrogen.

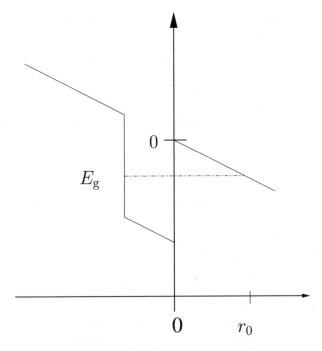

Fig. 4.3. Tunneling out of a square well supporting a bound state at E_g

Of central importance for the field of tunnel ionization is the definition of the Keldysh parameter

$$\gamma_K = \sqrt{\frac{I_p}{2U_p}}, \tag{4.18}$$

where I_p is the ionization potential of an atom and the ponderomotive potential U_p is given by (3.27). γ_K compares the strength of the external field with that of a typical atomic energy. In the case of hydrogen and in SI units $I_p \approx 13.6\,\text{eV}$ and at a wavelength of 800 nm and intensity of $10^{14}\,\text{W\,cm}^{-2}$ $\gamma_K \approx 1$. For $\gamma_K \gg 1$ the external field is only a small perturbation, whereas for $\gamma_K < 1$ the atomic potential plays the role of a perturbation and tunnel ionization starts to become the dominant mechanism.

The Keldysh or adiabaticity parameter can also be defined with the help of a "barrier time" as

$$\gamma_K \equiv 2\omega t_{ba}. \tag{4.19}$$

where the particle "spends" the time t_{ba} under the barrier. Furthermore, half the period of the external field is given by the external frequency via $1/(2\nu)$. In order for the particle to tunnel before the field turns around $\gamma_K < 2\pi$ and to be on the safe side even $\gamma_K < 1$ should be fulfilled. The length of the barrier in a potential as the one shown in Fig. 4.3 is

$$r_0 = \frac{-E_g}{\mathcal{E}_0} =: \frac{I_p}{\mathcal{E}_0}. \qquad (4.20)$$

The virial theorem tells us that the average kinetic energy of the electron is given by the ionization potential. We therefore assume that the electron tunnels through the barrier with a velocity of

$$v = \sqrt{2I_p}. \qquad (4.21)$$

The barrier time can then be estimated to be [11]

$$t_{ba} = \frac{r_0}{v} = \sqrt{\frac{I_p}{2\mathcal{E}_0^2}} \qquad (4.22)$$

and after inserting this into the Keldysh parameter

$$\gamma_K = \frac{\omega}{\mathcal{E}_0}\sqrt{2I_p} = \sqrt{\frac{I_p}{2U_p}} \qquad (4.23)$$

with $U_p = \mathcal{E}_0^2/(4\omega^2)$ follows. This consideration should be taken "with a grain of salt".

4.2.2 Multiphoton Ionization

Up to now we have considered the quasistatic case of a Coulomb potential distorted by a constant electric field. In the focus of the present book, however, is the question which effect a *time-dependent* external field exerts on an initially bound electron. If the energy of the photon is not enough to overcome the ionization threshold and if $\gamma_K \gg 1$, so that tunnel ionization is extremely unlikely then the transition to the continuum may happen via the absorption of several photons. In this so-called multiphoton ionization (MPI) two principally different scenarios are usually distinguished:

- *Resonant MPI (REMPI).* Here the spectrum of the unperturbed system is related to the photon frequency, a review of the topic can be found in [12]
- *Nonresonant MPI.* The frequency of the photons is not in resonance with energy differences of the system

A simple model, that allows to understand the basic features of MPI is that of an electron in a Gauß potential given by

$$V_G(x) \equiv -V_0 \exp\{-\sigma x^2\}. \qquad (4.24)$$

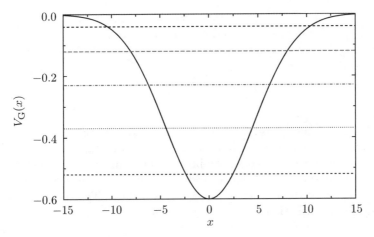

Fig. 4.4. Gauß potential (*solid line*) with eigenvalues (*horizontal lines*) for the parameters given in the text

This choice allows the investigation of resonance phenomena, that occur whenever the average level spacing is equal to the external frequency. Furthermore, the potential has the nice feature that, in contrast to the Coulomb case, it allows only for a finite number of bound states and therefore, the ionization probability can be calculated with great ease [13].

For the parameters $V_0 = 0.6$ a.u. and $\sigma = 0.025$ a.u. the potential together with its bound state energies at $E_0 = -0.518$, $E_1 = -0.37$, $E_2 = -0.23$, $E_3 = -0.12$, $E_4 = -0.04$, $E_5 = 0$ a.u. is depicted in Fig. 4.4.[4] The average distance between nearest neighbors is

$$\Delta E = \frac{1}{4} \sum_{i=0}^{3} (E_{i+1} - E_i) \approx 0.12 \text{ a.u.} \tag{4.25}$$

Furthermore, by using WKB quantization

$$W(E) = \oint p \, dx = (n + 1/2) 2\pi \tag{4.26}$$

with the short action $W(E) = S(t) + Et$, the spectrum is very well represented [14].

In order to study ionization, the initial state is taken as the ground state of the system, which is to a good approximation given by a Gaussian wavepacket

$$\Psi_0(x) = \Psi_\gamma(x) = \left(\frac{\gamma}{\pi}\right)^{1/4} \exp\left\{-\frac{\gamma}{2} x^2\right\} \tag{4.27}$$

---

[4] Note that each symmetric well potential in 1d has at least one bound state, no matter how shallow it is [9].

with $\gamma = 0.154$ a.u. The ionization probability under laser irradiation as a function of time in the case of the Gauss potential can be calculated most easily from the finite(!) sum

$$P_I(t) = 1 - \sum_{n=0}^{5} |\langle \Psi(t) | \psi_n \rangle|^2. \tag{4.28}$$

This is the probability *not* to be in a bound state any more. This probability typically increases as a function of time, if the electron can climb up the ladder of energy eigenstates and leave the bound states. In order to calculate it, the solution of the time-dependent Schrödinger equation is needed, and therefore we need to know the explicit form of the laser field. The one that has been used in the investigations of van de Sand [14] is

$$V_L = \mathcal{E}_0 x \sin^2 \left(\frac{\omega t}{16} \right) \sin(\omega t), \tag{4.29}$$

where the first sine factor is the envelope of the pulse, extending over eight cycles, $T = 2\pi/\omega$, of the field (thereafter the pulse is assumed to be zero). The field strength is $\mathcal{E}_0 = 0.032$ a.u. The frequency of the laser has been varied in order to study resonance phenomena. The numerical results have been gained by solving the full quantum problem (using the split-operator FFT method) as well as by using the semiclassical Herman–Kluk propagator.

(a) Nonresonant case: $\omega = 0.09$ a.u.

In Fig. 4.5, the ionization probability as well as the occupation probabilities of the states 0–2 are depicted as a function of time. A Nonresonant Rabi oscillation (with amplitude much smaller than unity) between the ground and the first excited state can be observed. The fast oscillations in the occupation probabilities are due to the counter-rotating term, which would not be present if the RWA had been invoked.[5] Furthermore, some probability leaks into the continuum and at the end of the pulse the ionization probability is about 15%. The semiclassical results in panel (b) very nicely reproduce the quantum ones.

(b) Resonance case: $\omega = 0.124$ a.u.

For the results shown in Fig. 4.6, the same field strength as above has been used. The frequency, however, is now close to the average level spacing. The final ionization then is around 80% and the wavepacket methodology is very well suited to describe this nonperturbative effect of the laser field. From the figure it can be seen that the occupation of the ground state decreases in a quasimonotonous way (and does not rise at the end of the pulse as in case (a)). Probability is being transferred successively into

---

[5] A direct comparison with the RWA results of Chap. 3 would ask for an extension of the two-level results to pulsed driving and in addition the dipole matrix element would have to be calculated.

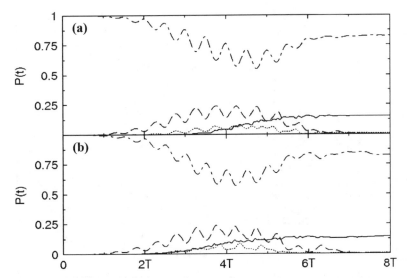

Fig. 4.5. Ionization probability (*solid line*) and occupation probability ($n = 0$ *long dash–short dash*, $n = 1$ *dashed*, $n = 2$ *dotted*) as a function of time in the nonresonant case $\omega = 0.09$ a.u. (**a**) full quantum results, (**b**) semiclassical results; adapted from [14]

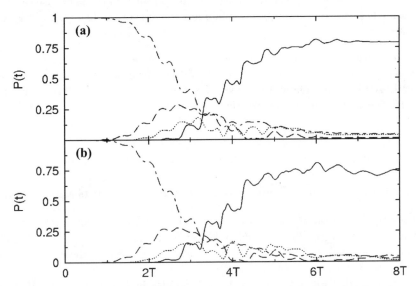

Fig. 4.6. Ionization probability (*solid line*) and occupation probability ($n = 0$ *long dash–short dash*, $n = 1$ *dashed*, $n = 2$ *dotted*) as a function of time in the resonant case $\omega = 0.124$ a.u. (**a**) full quantum results, (**b**) semiclassical results; adapted from [14]

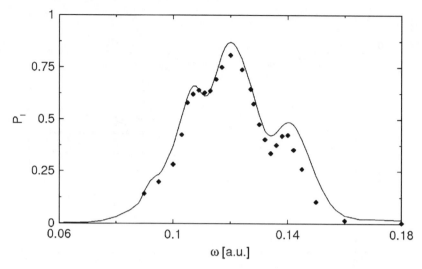

Fig. 4.7. Ionization probability P_I as a function of the laser frequency, semiclassical results are depicted by the *filled diamonds*; from [14]

higher states and finally into the continuum $E > 0$. This process is called REMPI (Resonantly Enhanced Multiphoton Ionisation). As in case (a), the semiclassical results reproduce the main features of the resonant dynamics very well.

The ionization probability after the laser excitation as a function of external frequency is given in Fig. 4.7. Maxima in $P_I(8T)$ occur at $\omega = 0.12$ a.u. and at $\omega = 0.14, 0.107$ a.u.. These are again very well reproduced semiclassically and are related to transitions in the spectrum as follows

$$n = 2 \rightarrow n = 3 : \qquad \omega = 0.107 \, \text{a.u.} \tag{4.30}$$

and

$$n = 0 \rightarrow n = 1 \quad \text{and} \quad n = 1 \rightarrow n = 2 : \qquad \omega = 0.14 \, \text{a.u.} \tag{4.31}$$

In principal, even higher ionization probabilities could be achieved if the frequency of the external field would be allowed to change in the course of time. So-called down-chirps have been investigated in molecular dissociation and will be dealt with in detail in Chap. 5.

To finish this section, a short remark on classical trajectory calculations to reproduce the quantum results is in order. These calculations are based on the solution of Hamilton's equations and do *not* take phase information into account. They have been performed by van de Sand, and for the present problem lead to considerably worse results than semiclassics [14].

4.2.3 ATI in the Coulomb Potential

If the photon frequency is of the order of an atomic unit, then already the absorption of a single photon is enough to ionize an atom (the ground state energy of hydrogen is -0.5 a.u.). Intriguingly, even in that case an atom can absorb more than one single photon, however, as we will see in the following. This breed of MPI is called above threshold ionization (ATI). Experimentally, the effect has first been shown to exist in the end of 1970s by using laser driven Xe atoms [15]. Theoretical studies showing the effect have been done for several models of the hydrogen atom. In the following a study using both 1d models, the soft-core and also the bare Coulomb potential from Sect. 4.1.2 will be reviewed.

For the numerical study of ATI an external field of the form

$$\mathcal{E}(t) = \mathcal{E}_0 f(t) \sin(\omega t) \tag{4.32}$$

$$f(t) = \begin{cases} \sin^2[\pi t/(2\tau_1)] & 0 < t < \tau_1 \\ 1 & \tau_1 \leq t \leq \tau_2 - \tau_1 \\ \cos^2[\pi(t + \tau_1 - \tau_2)/(2\tau_1)] & \tau_2 - \tau_1 < t < \tau_2 \end{cases} \tag{4.33}$$

with the parameters

$$\mathcal{E}_0 = 1 \text{ a.u.} \tag{4.34}$$

$$\omega = 1 \text{ a.u.} \tag{4.35}$$

has been used in [5]. The ponderomotive potential, respectively, the Keldysh parameter then have the values $U_p = 0.25$ a.u., $\gamma_K = 1$. The total pulse duration is 25 optical cycles ($\tau_2 = 50\pi/\omega$) and the field is switched on over five optical cycles ($\tau_1 = 10\pi/\omega$), is constant for 15 optical cycles, and is switched off in another five optical cycles.

Numerical results from [5] for the initial condition $\Psi(x,0) = \Psi_1^{\sigma=1}(x)$ are depicted in Fig. 4.8 for the bare and the soft-core (with $a = 1$) Coulomb potential. Especially in the case of the bare Coulomb potential a distinct splitting of the wavepacket into different subpackets can be observed. These subpackets correspond to parts of the wavefunction, that have absorbed a different number of quanta of radiation energy. Although the field is treated classically, the excitation of the system shows clear peaks around multiple integers of the photon energy, which leads us to speak of the absorption of quanta of energy. There is a certain probability that an electron has asymptotically the velocity $\Delta x/\Delta t$ corresponding to a kinetic energy equal to the energy of the ATI peak. For the present field parameters and for the hard-core potential, the first peak in the spectrum is at an energy of $1/2$ a.u.. This is the Einstein-peak because it is the peak well known from the photoelectric effect, for the explanation of which Einstein was awarded the 1921 Nobel Prize in Physics. Further ATI peaks well visible in the bare Coulomb case are separated by 1 a.u.

Comparing the results for the bare and the soft-core Coulomb case in Fig. 4.8, we first note that the Einstein-peak is shifted toward lower energies

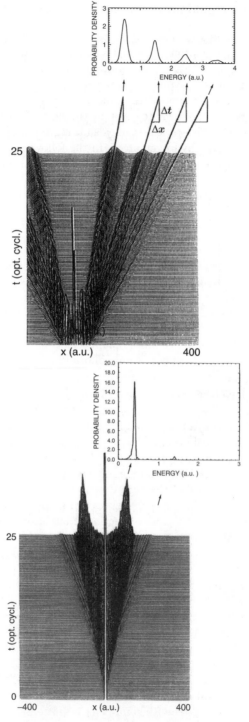

Fig. 4.8. Photoelectron spectra and wavefunctions for ATI in the "bare" (*upper panels*) and in the "soft-core" (*lower panels*) Coulomb-Potential, adapted from [5]

in the second case. This is due to the fact that $a = 1$ has been chosen for the soft-core parameter here. For $a = 2$ the peak would be at the same position as in the bare Coulomb case. More importantly, however, the soft-core potential leads to a dramatic underestimation of the first as well as all higher ATI peaks! A theoretical understanding of the effect can be gained by using the Herman–Kluk propagator as has been shown by van de Sand [14]. It turns out that interfering trajectories are responsible for the formation of the ATI peaks, which cannot be described by using pure classical mechanics.

4.2.4 Stabilization in Very Strong Fields

A counterintuitive phenomenon has been predicted theoretically to occur in very strong external fields, i.e., fields that are even stronger than an atomic unit. It has been shown that a continued increase of the field strength will eventually lead to a decrease of photoionization. Frequently used techniques to study this phenomenon are the strong field approximation or the Keldysh–Faisal–Reiss approximation. A review of these methods and their differences can be found in [16].

Due to the fact that most of the features of stabilization can be understood classically, in the following we focus on a classical mechanics study of a 1d soft-core Coulomb potential, driven by a monochromatic electric field

$$\mathcal{E}(t) = \mathcal{E}_0 f(t) \sin(\omega t). \tag{4.36}$$

If the field is switched on over 5 periods ($\omega = 0.8$) and then oscillates for 50 periods with the constant amplitude $\mathcal{E}_0$, the behavior shown in Fig. 4.9 is observed [17]. For this figure 5,000 different bound ($E < 0$) initial conditions have been chosen and the solution of Newton's equation has been calculated numerically. As a function of the field amplitude, the fraction of electrons that have positive energy after the pulse (i.e., the fraction of ionized electrons) is then plotted. For fields larger than an atomic unit, ionization as a function of field strength on average drops monotonically.

This phenomenon, which is referred to as stabilization against ionization, can be explained by going to the Kramers–Henneberger frame of Chap. 3. The classical Hamiltonian of the driven system then is given by

$$H = \frac{p^2}{2} - \frac{1}{\sqrt{1 + [q + \alpha(t)]^2}}, \tag{4.37}$$

where in atomic units (and in 1d)

$$\dot{\alpha}(t) = -A(t), \tag{4.38}$$

and thus

$$\ddot{\alpha}(t) = \mathcal{E}(t) \tag{4.39}$$

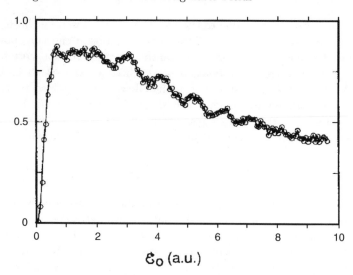

Fig. 4.9. Fraction of ionized electrons in an almost monochromatic laser pulse as a function of field strength, from [17]

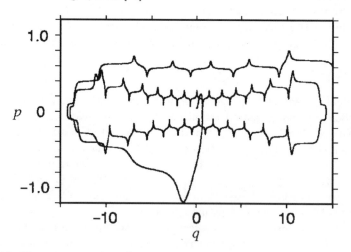

Fig. 4.10. Phase space orbit of an electron for $\omega = 0.8$, $\alpha_0 = 11.7$, and $q(0) = p(0) = 0$ adapted from [17]

hold. The solution of this differential equation after the field is switched on is given by

$$\alpha(t) = -\alpha_0 \sin(\omega t), \qquad (4.40)$$

with the quiver amplitude $\alpha_0 = \mathcal{E}_0/\omega^2$. The nucleus oscillates with a typical velocity of $\omega \alpha_0$ between the turning points $\pm \alpha_0$.

For the electron, moving in the field of the nucleus, the behavior shown in Fig. 4.10 is typically found. Starting from the origin of phase space, after an initial acceleration, the electron is drifting into a certain direction for a long time and is passed by the nucleus frequently (twice in an optical cycle). At each encounter it is shortly pulled either to the left or to the right (or the other way round). The net effect is very small because the effect of the two encounters almost cancel each other. If the nucleus is at a turning point, however, its influence on the electron motion can be very strong. First of all, the nucleus is very slow at the turning point and can act for a long time on the electron and secondly, both encounters can have the same effect, because the nuclear motion may have changed its direction in-between both events. The electron then also turns around [17] and acquires energy until it has enough to leave the nucleus.

Stabilization can now be explained by the fact that for strong fields, the electron stays for longer and longer times in regions where no energy gain is possible. These regions increase in size with growing α_0.

4.2.5 Atoms Driven by HCP

The external driving considered so far was either static or (almost) monochromatic. What happens if a pulse, which is very short and therefore very polychromatic, hits an atomic system? The ultimate form of such a short pulse is a HCP. It is assumed to be nonzero only for half an oscillation period. The generation of such pulses is the subject of a book on its own. Depending on the shortness required, they can be generated by irradiating semiconductors with pulsed optical lasers [18, 19], or by applying voltage pulses on capacitor electrodes [20].

We begin this section by considering a hydrogen atom in the field of a single pulse and then switch to the case of a Rydberg atom in a periodic train of HCP.

Ground State Hydrogen Under a Single HCP

A hydrogen atom coupled to an electric field in length gauge is governed by the Hamiltonian

$$\hat{H}(t) = \hat{H}_0 + \hat{r} \cdot \mathcal{E}(t) \tag{4.41}$$

with the unperturbed 3d Coulomb Hamiltonian $\hat{H}_0$. How can the time-dependent Schrödinger equation be solved for HCP driving? The answer to this question is given by an application of the Magnus expansion of Sect. 2.2.4. When the length T_p of the pulse is very short compared to a typical orbital time, as e.g., $T_e = 2\pi$ in the case of ground state dynamics in the hydrogen atom, and the field amplitude is very high, the so-called first-order Magnus approximation

$$|\Psi(t)\rangle \approx \exp\left(-i \int_0^t dt' \hat{H}(t')\right) |\Psi(0)\rangle$$

$$\approx \exp\left[-i \left(\int_0^t dt' \mathcal{E}(t')\right) \cdot \hat{r}\right] \exp[-i\hat{H}_0 t]|\Psi(0)\rangle \tag{4.42}$$

for the wavefunction can be used [21]. Analogous to the procedure in the interaction picture presented in Sect. 2.2.4, it can be derived from the exact time-evolution operator in the Schrödinger picture by neglecting the time-ordering operator. In addition, also the noncommutativity of the perturbation with the unperturbed Hamiltonian is neglected. Both approximations improve, the shorter the time span of the perturbation is.

The first exponent in the second line of the solution above leads to the definition of the momentum that is transferred to the atomic system

$$\Delta \boldsymbol{p} := - \int_0^t \mathrm{d}t' \boldsymbol{\mathcal{E}}(t'). \tag{4.43}$$

If the initial state is an eigenstate of $\hat{H}_0$, then apart from the phase due to the application of the unperturbed Hamiltonian, the momentum change is the main effect of the total Hamiltonian. Due to the fact that only the pulse area[6] appears in the expression above, the HCP could also have been assumed to be a δ-pulse with the appropriate strength

$$\boldsymbol{\mathcal{E}}(t) = -\Delta \boldsymbol{p} \delta(t - t_1), \qquad t_1 \in (0, T_{\mathrm{p}}). \tag{4.44}$$

In Fig. 4.11, taken from [21], it is shown that in the case of the hydrogen atom starting from the ground state, for a HCP with $T_{\mathrm{p}} = 0.3$ a.u. and with the absolute value of the momentum transfer $q = -\Delta p > 3$ the atom is ionized with almost certainty. Furthermore, full DVR calculations show that no matter if the pulse is rectangular, with $\mathcal{E}_0 = q/T_{\mathrm{p}}$ or sinusoidal, with $\mathcal{E}_0 = \pi q/(2T_{\mathrm{p}})$, the outcome is identical to the first-order Magnus results. This section shall be closed by a short remark on the orders of magnitude that are needed to experimentally realize the system studied above. In order for the first-order Magnus approximation to be applicable $T_{\mathrm{p}} < 1$ a.u. and thus the pulse length has to be of the order of attoseconds (10^{-18} s). Furthermore, the momentum transfer for total ionization has to be larger than one atomic unit and therefore, the corresponding field strengths have to be of the order of several atomic units and the intensities should be of the order of 10^{18} W cm^{-2}.

Rydberg States in Periodic Trains of HCP

If the initial state with energy E_n is a highly excited Rydberg state with a relatively long Kepler period of

$$T_n = T_{\mathrm{e}} n^3 = 2\pi n^3, \tag{4.45}$$

then the absolute length of the HCP may be very much longer than in the case studied in the previous section and can still be considered to deliver a δ-function like kick to the atom. For Rydberg states of principal quantum

---

[6] This is another occurrence of the area theorem of Sect. 3.2.

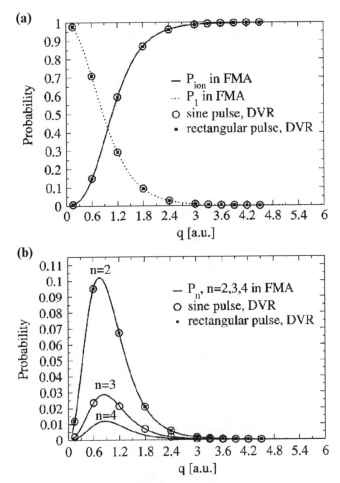

Fig. 4.11. (a) Ionization probability and occupation probability of the ground state, (b) some occupation probabilities for excited states of the hydrogen atom in very intense HCPs as a function of momentum transfer q [21]

number $n = 400$, Kepler periods are in the range of nanoseconds and for $n = 60$ they are in the picosecond range.

An investigation of the effects of a single pulse on the dynamics of Rydberg atoms has been done in [22]. Many interesting effects do emerge, however, if a periodic train of HCPs with period T is applied to a Rydberg atom. A combined experimental and theoretical study of such a system is reported in [20].

In the following, we consider in some detail the bare 1d Coulomb potential with a periodic kicking term that leads to the classical Hamiltonian

$$H^{1d}(p, q, t) = \frac{p^2}{2} - \frac{1}{q} - q\Delta p \sum_{k=1}^{N} \delta(t - k/\nu_T) \qquad (4.46)$$

in terms of phase space variables (q, p) with $\nu_T = 1/T$. Here we also assume that the motion is restricted to the positive half space which can be motivated by the presence of a centrifugal barrier due to a quasiangular momentum Λ, which we let go to zero [20], and which can also be realized experimentally by photo excitation of selected Stark states [23]. Due to the instantaneous kicking the stroboscopic dynamics (over one period) can be factorized according to

$$(q_k, p_k) = M_{\text{Coul}} M_{\Delta p}(q_{k-1}, p_{k-1}) \tag{4.47}$$

into the kick contribution and an unperturbed Coulomb term.

As in quantum mechanics, the kicking term's action is to change the momentum according to

$$(q_{k-1}, p_{k-1} + \Delta p) = M_{\Delta p}(q_{k-1}, p_{k-1}), \tag{4.48}$$

leading to the new energy

$$E_k = \frac{\Delta p^2}{2} + p_{k-1}\Delta p + E_{k-1}. \tag{4.49}$$

The energy conserving Coulomb dynamics over one period of the driving can be extracted from the parametric form of the orbit, which for negative energies is given by [24]

$$q = n_k^2(1 - \epsilon \cos \xi), \qquad \xi - \epsilon \sin \xi = n_k^{-3} t \tag{4.50}$$

with the eccentricity ϵ, which for vanishing quasiangular momentum considered here is equal to unity, and $n_k = (2|E_k|)^{-1/2}$. Due to the fact that the Coulomb potential is depending only on a single power of the position variable, the scaling transformations with the initial quantum number n

$$q = n^2 q_0, \tag{4.51}$$
$$p = n^{-1} p_0, \tag{4.52}$$
$$t = 2\pi n^3 t_0 \tag{4.53}$$

can be used to extract the factor n^{-2} from the Hamiltonian. Analogously, for the frequency

$$\nu = \nu_0/(2\pi n^3) \tag{4.54}$$

applies.

Two different types of classical dynamics can be distinguished. For $\Delta p_0 > 0$, i.e., a kicking force that acts away from the center, one finds that even for very weak driving the phase space plot is completely chaotic, see e.g., Fig. 4.12a. For a force that acts toward the center of Coulomb attraction ($\Delta p_0 < 0$), however, classical dynamics with a mixed phase space emerges, which is depicted in Fig. 4.12b, where regular and chaotic regions are both present.

Let us concentrate on kicks toward the singularity ($\Delta p_0 < 0$) in the following. The scaled frequency can be chosen either such that the phase space

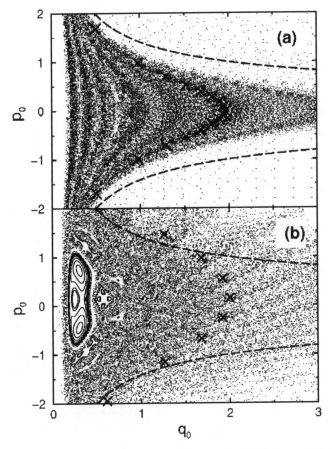

Fig. 4.12. Stroboscopic phase space plot with $n = 50$ and (a) $\Delta p_0 = 0.01, \nu_0 = 16.8$ and (b) $\Delta p_0 = -0.3, \nu_0 = 15.9$. The *crosses* are periodic orbits and the *dashed line* shows the location of the ionization threshold energy $E = 0$ [25]

contour for the initial energy is in a regular area (see e.g., the left Poincaré section in Fig. 4.13, where the microcanonical initial state is indicated by a gray line), or it can be chosen such that the gray line is in a chaotic area (see the Poincaré section on the right). In the first case by comparing classical and quantum ionization dynamics, a stabilization (or localization) of the classical results for the survival probability

$$P_s(t) = \begin{cases} \sum_n |\langle \psi_n | \Psi(t) \rangle|^2 & \text{quantum} \\ \int_{E<0} \mathrm{d}p\,\mathrm{d}q\, f(q, p, t) & \text{classical} \end{cases} \tag{4.55}$$

is found (here f is the phase space function corresponding to the quantum mechanical wavefunction). In the second case, if the initial state is in the chaotic range, the opposite effect is observed. Now quantum mechanically the

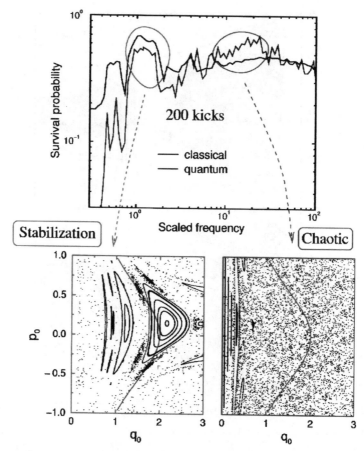

Fig. 4.13. Survival probability after 200 kicks as a function of the scaled frequency ($n = 60$) for $\Delta p_0 = -0.3$ (*upper panel*) and Poincaré sections for two different frequencies (*two lower panels*) [26]

results are localized. This effect remains if the number of kicks is increasing as can be seen in Fig. 4.14.

A semiclassical understanding of the observed effects would be a real breakthrough. First indication of semiclassical localization has been given in a study of the kicked rotor [27]. Right at the onset of localization it is becoming extremely hard, however, to converge the numerical results. In the case of HCP driven Rydberg atoms, as can be seen in Fig. 4.14, application of the Herman–Kluk propagator shows the localization up to around 100 kicks for the chosen parameters. For a larger number of kicks, the results tend toward the classical ones, however. Also here, convergence is a problem due to the fact that at longer times, the Herman–Kluk prefactors assume values of 10^{20} [25].

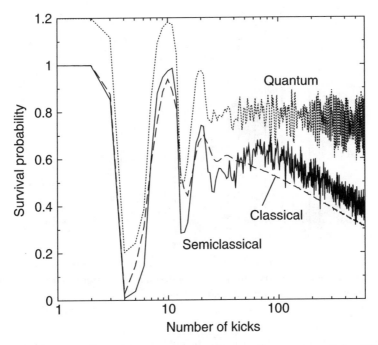

Fig. 4.14. Survival probability as a function of time. Quantum result is shifted by 0.2 for better visibility [25]

4.3 High Harmonic Generation

Apart from being ionized, an atom that is irradiated by laser light of a certain frequency ω can be radiating itself by emitting photons at odd harmonic frequencies $(2n + 1)\omega$ of the original radiation. A schematic representation of the effect is given in Fig. 4.15. High harmonic generation (HHG) via laser excitation has been demonstrated experimentally with harmonics as high as 500 times the fundamental frequency[7] having been reported, and it is reviewed in [28].

4.3.1 Three-Step Model

A simple explanation of HHG can be given by considering the following three-step model [30]:

- The laser field distorts the Coulomb potential, such that the electron can tunnel out of the range of attraction (see e.g., Fig. 4.2)
- The (almost) free electron is accelerated in the laser field

---

[7] The creation of radiation in the soft X-ray regime (with photon energies around 1 keV) is thus feasible [29].

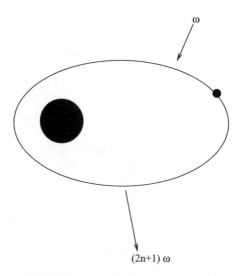

Fig. 4.15. Schematic of HHG for a single electron atom: After irradiation by a field of frequency ω, odd harmonics are emitted

- Upon return of the electron to the nucleus when the field turns around, the electron recombines with the ion and the energy is released in the form of a single photon

It is important to note that the answer of the electron to the applied field is not instantaneous as in the case of low field nonlinear optics but there is a time delay between the stimulus and the radiation of the high harmonics.

In order to describe HHG also quantitatively, a measurable quantity has to be defined. This is the harmonic spectrum which can be expressed in terms of the Fourier transformation of the dipole acceleration given by

$$
\begin{aligned}
a(t) &= \frac{\mathrm{d}^2}{\mathrm{d}t^2}\langle \hat{x}\rangle(t) = \frac{\mathrm{d}^2}{\mathrm{d}t^2}\langle \Psi(t)|\hat{x}|\Psi(t)\rangle \\
&= \frac{\mathrm{d}}{\mathrm{d}t}\langle \Psi(t)|\hat{p}|\Psi(t)\rangle \\
&= -\langle \Psi(t)|\frac{\mathrm{d}V}{\mathrm{d}\hat{x}}|\Psi(t)\rangle.
\end{aligned}
\tag{4.56}
$$

In the equations above, the Ehrenfest theorem

$$
\mathrm{i}\frac{\mathrm{d}}{\mathrm{d}t}\langle \hat{A}\rangle = \langle [\hat{A},\hat{H}]\rangle + \mathrm{i}\left\langle \frac{\partial}{\partial t}\hat{A}\right\rangle
\tag{4.57}
$$

for the time-evolution of expectation values has been applied once for $\hat{A} = \hat{x}$ and once for $\hat{A} = \hat{p}$. The term due to the laser potential in (4.56) leads only to a contribution at the fundamental frequency and is therefore usually subtracted. In the expectation value then only the term $\frac{dV_C}{d\hat{x}}$ remains. The spectrum is finally given by the Fourier transformation

$$\sigma(\Omega) = \frac{1}{\sqrt{2\pi}} \int_0^{T_t} dt\, e^{-i\Omega t} \langle \ddot{\hat{x}} \rangle (t)$$

$$= \frac{1}{\sqrt{2\pi}} \left[e^{-i\Omega T_t} \langle \dot{\hat{x}} \rangle (T_t) + i\Omega e^{i\Omega T_t} \langle \hat{x} \rangle (T_t) - \Omega^2 \int_0^{T_t} dt\, e^{-i\Omega t} \langle \hat{x} \rangle \right]. \quad (4.58)$$

In going to the second line it has been assumed that $\langle \hat{x} \rangle (0) = \langle \dot{\hat{x}} \rangle (0) = 0$. In some early literature also the Fourier transformation of the dipole expectation value, i.e., only the last term in the second line of the equation above is used to calculate the harmonic spectrum. This is generally not correct, due to the fact that the boundary terms in the partial integration do not necessarily vanish [31]. The unraveling of the harmonic spectrum when the length T_t of the integration interval is increased is displayed in Fig. 4.16.

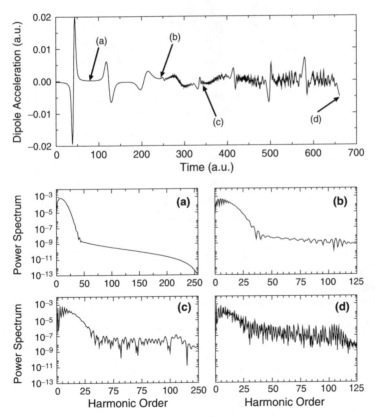

Fig. 4.16. Features of HHG emerging by increasing the time interval T_t, from [32]

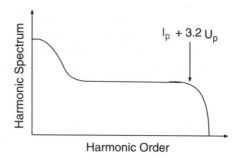

Fig. 4.17. Schematic of the HHG spectrum: After an initial decay and a long plateau, the spectrum drops quickly at a cutoff value of the harmonic order

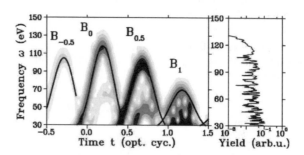

Fig. 4.18. Husimi plot of the dipole acceleration, simulating an experiment with Ne in a 5-fs laser pulse at 800 nm wavelength and with an intensity of $5 \times 10^{14}\,\mathrm{W\,cm^{-2}}$, from [33]

As expected by using perturbation theory, the intensity of the harmonics initially decreases with higher order n. Unexpectedly, however, a long *plateau* region can be observed in Fig. 4.16, which is ending with a sharp so-called *cutoff*. These main features are once more depicted schematically in Fig. 4.17. The following questions regarding HHG await an answer:

(a) Why is there a sharp cutoff?
(b) Why are harmonics only observed at *odd* multiples of the fundamental frequency?
(c) Why is there a long plateau in the intensity of the harmonics?

By a simple argument using classical mechanics, the answer to question (a) can be given. The explanation of questions (b) and (c) are a little bit more involved and will therefore be dealt with in extra sections.

Before explaining the different features, a further remark on the noninstantaneous nature of HHG shall be made. It has been shown by numerical simulations of an experiment with Ne atoms that the highest harmonics are created in very short time intervals of the length of attoseconds [33]. Figure 4.18, for

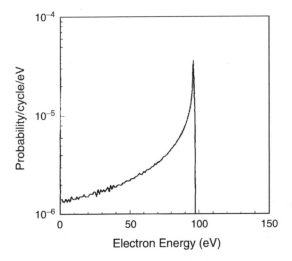

Fig. 4.19. Energy distribution of electrons at the first encounter with the ion in the case of Helium and with $I = 5 \times 10^{14}\,\mathrm{W\,cm^{-2}}$, $\lambda = 800\,\mathrm{nm}$; from [30]

example, shows the Husimi transform, defined in Chap. 1, of the dipole acceleration. This observation also opens a road to the generation of attosecond pulses of electromagnetic radiation.

The Cutoff

The cutoff can be explained by using the three-step model. The decisive question is how much energy an electron can acquire from a monochromatic laser field. To answer this question, numerical studies based on classical mechanics have been performed [30]. The outcome of these calculations is depicted in Fig. 4.19. It is shown there that an electron that is initially bound in an atom can acquire an energy of maximally $3.17U_\mathrm{p}$ with the ponderomotive potential from (3.27). The maximal energy that can be released upon recombination with the ion is therefore given by $E_\mathrm{C} \approx 3.17U_\mathrm{p} + I_\mathrm{p}$ with the ionization potential I_p. This explains the sharp cutoff observed in the harmonic spectrum.[8]

4.3.2 Odd Harmonics Rule

To explain the peaks of the HHG spectrum at odd harmonics we have to invoke some quantum mechanical reasoning. In the case of a periodically driven quantum system, the Floquet theorem of Sect. 2.2.7 (in a.u.)

---

[8] In the laser field only, as we saw in Chap. 3, the maximal energy that can be gained is $2U_\mathrm{p}$, and the cutoff for initially free electrons has to be adjusted accordingly as we will see later.

$$\Psi_n(x,t) = \exp\left\{-i\epsilon_n t\right\} \psi_n(x,t), \tag{4.59}$$

$$\psi_n(x,t) = \psi_n(x,t+T) \tag{4.60}$$

holds. In the case of a symmetric potential, due to symmetry under the generalized parity transformation

$$\mathcal{P}: x \rightarrow -x, \qquad t \rightarrow t + T/2, \tag{4.61}$$

reviewed in Appendix 3.A, one can conclude that the Floquet functions have to obey the additional condition

$$\psi_n(-x, t+T/2) = \pm\psi_n(x,t). \tag{4.62}$$

Using this fact and by considering the Fourier transformation of the dipole expectation value $\langle\psi_n|\hat{x}|\psi_n\rangle$ it can be shown (neglecting the boundary terms in (4.58)) that the HHG spectrum contains only odd harmonics.

Exercise 4.2 *Show that the dipole expectation value between Floquet states has only odd Fourier components.*

In the previous as well as the following numerical results, the theorem of odd harmonics is not fulfilled exactly, however. This is due to the fact that a pulsed driving laser is used and therefore, Floquet theory cannot be applied strictly.

Furthermore, the odd harmonics rule can be violated in *molecules* also due to the breakdown of the Born–Oppenheimer approximation, as has been shown in [34]. The Born–Oppenheimer or adiabatic approximation will be discussed in detail in Chap. 5.

4.3.3 Semiclassical Explanation of the Plateau

Understanding the plateau formation in HHG is a formidable task and has been undertaken by Gerd van de Sand during his PhD thesis work [14,35]. The main working horses used in these studies are the semiclassical Herman–Kluk propagator and a comparison with full quantum as well as purely classical calculations.

In order to perform all three types of calculation the Hamiltonian

$$H = \frac{p^2}{2} + V_{\text{sc}} + V_{\text{L}} \tag{4.63}$$

with the 1d soft-core Coulomb potential (4.15) and the laser potential in length gauge

$$V_{\text{L}} = f(t)\mathcal{E}_0 x \cos(\omega t) \tag{4.64}$$

has been used in [14]. The parameters appearing in the Hamiltonian have been chosen as follows:

- soft-core parameter $a = 2$ leading to an ionization potential of $I_p = 0.5$ equal to that of 3d hydrogen
- $\mathcal{E}_0 = 0.1$ a.u. and $\omega = 0.0378$ a.u., where $f(t)$ has been chosen such that the laser pulse lasts for 3.5 cycles of the oscillation

The initial wavefunction was assumed to be a Gaussian wavepacket

$$\Psi(x) = \left(\frac{\gamma}{\pi}\right)^{1/4} \exp\left\{-\frac{\gamma}{2}(x - q_0)^2\right\} \tag{4.65}$$

with $\gamma = 0.05$ a.u. and $q_0 = 70$ a.u., corresponding to an almost free initial electron being scattered as soon as it reaches the ion under the influence of the laser. This initial state has been chosen due to the fact that the initial tunneling step cannot be described semiclassically, and as we will see below it leads to a cutoff energy that is slightly reduced compared to that of the case of a bound initial state.

The dipole acceleration as well as the corresponding spectra are compared in Figs. 4.20, 4.21, and 4.22. The first figure shows the results of a full quantum mechanical calculation, the second those of a classical "Monte-Carlo" calculation [36], and the third plot shows semiclassical Herman–Kluk results. In all three cases, the dipole acceleration has maxima whenever the electron is close to the nucleus. Due to the broadening of the wavepacket in the course of time the maxima get smeared out. In the quantum, as well as in the semiclassical result, at longer times, fast oscillations of $a(t)$ occur, which are not present in the classical result.

The HHG spectra extracted from the time-domain results are displayed in the lower panels of the respective plots. The full quantum spectrum displays the plateau and a cutoff at 105 times the fundamental frequency. The energy corresponding to this frequency is close to $E_C = 2U_p + I_p$, which is the predicted cutoff for a scattering initial condition, differing slightly from the one for a bound state initial condition. The cutoff is well represented in the classical, as well as the semiclassical result. The plateau, however, can only be observed in the semiclassical case! The plateau formation therefore seems to be an interference effect.

In order to further investigate the nature of the interference effect, underlying the formation of the plateau, several different classes of classical trajectories have been identified. Apart from "almost free" trajectories, also stranded and trapped trajectories exist, as can be seen in Fig. 4.23. All trajectories in this figure start at the same initial position but with different initial momenta. Successively neglecting the trapped and stranded trajectories in the semiclassical calculation, the behavior in Fig. 4.24 can be observed: The plateau is vanishing. It is surprising that neglecting only around 8% of the trajectories is enough to completely suppress the formation of the plateau. These observations led the authors of [35] to conclude that the interference of stranded and trapped trajectories with the free ones is responsible for the plateau formation.

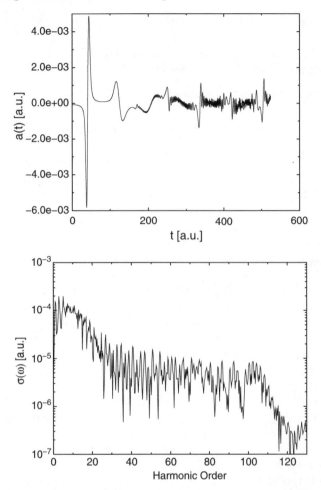

Fig. 4.20. Full quantum mechanical dipole acceleration and corresponding spectrum; adapted from [14]

4.3.4 Cutoff and Odd Harmonics Revisited

Apart from the explanation of the plateau, also the odd harmonics rule and the cutoff can be understood semiclassically. This is another application of the stationary phase approximation of Sect. 2.2.1 and shall dealt with below.

Motivated by the fact that the HHG can be explained as an interference effect of two classes of trajectories (almost free and trapped or stranded), the total wavefunction shall be composed of two parts [14]:

- An "ionization wavefunction", for the free electron in the laser potential V_L with $f(t) = 1$, which is given by the Volkov solution (3.24)

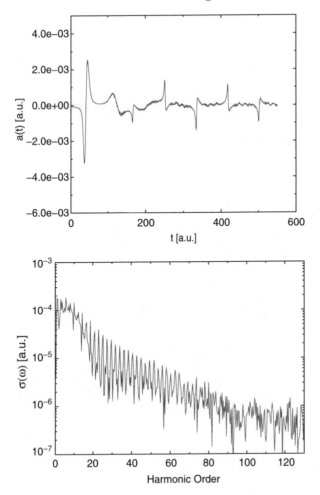

Fig. 4.21. Classical dipole acceleration and corresponding spectrum; adapted from [14]

- A "ground state wavefunction", of the electron in the soft-core Coulomb potential without the laser. The phase that this wavefunction acquires in the course of time is determined by the ionization potential via

$$\Phi(t) \approx I_p t. \tag{4.66}$$

It has been shown in [14] that a small admixture of the ground state wavefunction to the Volkov wavepacket leads to a close qualitative agreement with the exact results for the expectation value of the dipole acceleration.

In order to perform the SPA, the phase in the Fourier integral for the calculation of the HHG spectrum has to be known. This can be extracted

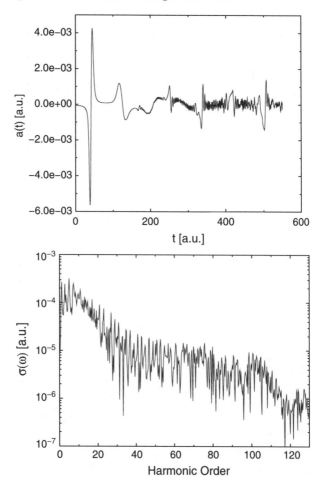

Fig. 4.22. Semiclassical dipole acceleration and corresponding spectrum; adapted from [14]

from the superposition[9] of the two wavefunctions leading to an expression of the form

$$\sigma(\Omega) = \int dt\, e^{i\Omega t} \left| \exp\left\{ i \left[\frac{U_p}{2\omega} \sin(2\omega t) - U_p t \right] \right\} + \exp\left\{ i I_p t \right\} \right|^2 \quad (4.67)$$

for the spectrum. Calculating the Fourier integral in the stationary phase approximation leads to the condition

$$\frac{d}{dt}\left[\Omega t \pm \left(\frac{U_p}{2\omega} \sin(2\omega t) - U_p t - I_p t \right) \right] = 0 \quad (4.68)$$

---

[9] For simplicity, we can take a superposition with equal weights for the following purposes.

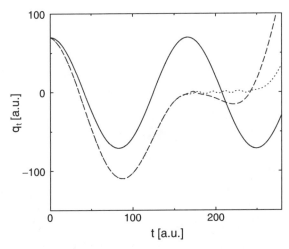

Fig. 4.23. "Almost free" (*solid line*) and "trapped" (*dashed line*) and "stranded" (*dotted line*) trajectories; adapted from [14]

and therefore the main contributions to the integral are at the (positive) frequencies

$$\Omega = U_p[1 - \cos(2\omega t)] + I_p. \tag{4.69}$$

The maximal value of the expression above follows immediately to be

$$\Omega_{\max} = 2U_p + I_p. \tag{4.70}$$

As in the discussion of the three-step model this frequency is generated by an electron that acquires the maximal energy in the laser field and transforms it into a single photon in an inverse photoelectric effect.

Also the fact that only odd harmonics are emitted can be derived by using the stationary phase approximation. The interference term in (4.67)

$$\int dt \exp\left\{ i \left[(\Omega - U_p - I_p) t + \frac{U_p}{2\omega} \sin(2\omega t) \right] \right\} \tag{4.71}$$

has to be studied to this end. If the exponent of the sine function is expanded in terms of Bessel functions using the Jacobi–Anger formula

$$\exp[i\alpha \sin(x)] = \sum_{n=-\infty}^{\infty} J_n(\alpha) \exp(inx), \tag{4.72}$$

in the exponent of (4.71) this leads to the replacement of the sine by its argument according to

$$\sum_n J_n \left(\frac{U_p}{2\omega} \right) \int dt \exp\left\{ i\left(\Omega - U_p - I_p + 2n\omega \right) t \right\}. \tag{4.73}$$

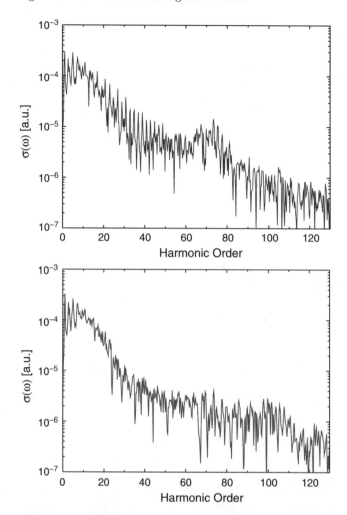

Fig. 4.24. Spectrum without trapped (*upper panel*) and without trapped and stranded trajectories (*lower panel*); adapted from [14]

Along the lines of the SPA, the maxima of this expression occur at

$$\Omega_n = U_{\rm p} + I_{\rm p} - 2n\omega \tag{4.74}$$

and are separated by an *even* number of fundamental frequencies 2ω. The fundamental is contained in the spectrum and therefore all the maxima are at *odd* harmonics.

The cutoff can once more be derived by looking at the asymptotic form

$$|n| \gg 1 : \qquad J_n(x) \to \frac{(-1)^n}{\sqrt{2\pi|n|}} \left(\frac{ex}{2|n|} \right)^{|n|} \tag{4.75}$$

of the Bessel function (here e = 2.72, Euler number). For $|n| > (e/2)\frac{U_p}{2\omega} \approx \frac{U_p}{2\omega}$ the Bessel function goes to zero, leading to an upper bound $\hbar\Omega_{\max} = 2U_p + I_p$, if for n the minimal value $-U_p/2\omega$ is inserted in (4.74).

4.A More on Atomic Units

There is a close connection between atomic units and some expectation values of the hydrogen atom. The Bohr radius, e.g., is related to the expectation value of the position operator in the ground state which is given by

$$\langle r \rangle = \frac{3}{2} a_0, \tag{4.76}$$

and in addition the maximum of the probability density for finding the particle in the ground state with a radial separation r from the nucleus is a_0 [2]. Furthermore, the energy constructed from the atomic base units is twice the absolute value of the ground state energy $E_1 = -\frac{1}{2}$ a.u.$\approx -13.6\,\mathrm{eV}$. One could also measure energy in terms of the ionization potential

$$I_p = -E_1 \tag{4.77}$$

of hydrogen, however. The corresponding units are so-called Rydberg units (1 a.u.=2 Ry), in contrast to the atomic or Hartree units.

The atomic unit for velocity follows most easily by considering the virial theorem, which for the hydrogen atom can be stated as

$$\langle T \rangle = -\frac{1}{2} \langle V_C \rangle. \tag{4.78}$$

For the ground state we find with its help that

$$\langle T \rangle = -E_1 \tag{4.79}$$

and with the definition

$$\langle T \rangle = \frac{m_e v_0^2}{2} \tag{4.80}$$

the atomic velocity unit $v_0 = \sqrt{2|E_1|/m_e} \approx c\sqrt{27.212\,\mathrm{eV}/0.511\,\mathrm{MeV}}$ is related to the vacuum speed of light via $v_0 = c\alpha = 1$ a.u. with the fine structure constant $\alpha \approx 1/137$. Using v_0 and a_0 the atomic unit of time is formed via $t_0 = a_0/v_0$. The oscillation period of the electron on the Bohr orbit is therefore given by $T_e = 2\pi$ a.u.

The most important unit for this book is the one for the electric field which is given by $\mathcal{E}_{\mathrm{at}} \approx 5.1427 \times 10^{11}\,\mathrm{V\,m^{-1}}$. This is the value of the electric field, due to the proton, experienced by the electron at the Bohr radius. The intensity that corresponds to that field is $I_{\mathrm{at}} = \frac{c\varepsilon_0\mathcal{E}_{\mathrm{at}}^2}{2} \approx 3.5101 \times 10^{16}\,\mathrm{W\,cm^{-2}}$.

Table 4.1. Definition of atomic units and some important quantities in atomic and SI units

	SI (m, kg, s, A)	a.u. $(a_0, m_e, \hbar, e)$
Definitions		
$\hbar$	$\approx 1.0546 \times 10^{-34}$ J s	$= 1\,\hbar$
e	$\approx 1.6022 \times 10^{-19}$ A s	$= 1\,e$
m_e	$\approx 9.1094 \times 10^{-31}$ kg	$= 1\,m_e$
a_0	$\approx 5.2918 \times 10^{-11}$ m	$= 1\,a_0$
Further quantities		
$4\pi\varepsilon_0 = \frac{10^7}{c^2}\,\mathrm{A\,m\,(V\,s)}^{-1}$	$\approx 1.1126 \times 10^{-10}\,\mathrm{A\,s\,(V\,m)}^{-1}$	$= 1\,\frac{a_0\,m_e\,e^2}{\hbar^2}$
$v_0 \approx \frac{1}{137.036}c$	$\approx 2.1877 \times 10^6\,\mathrm{m\,s}^{-1}$	$= 1\,\frac{\hbar}{a_0\,m_e}$
$2|E_1| = \frac{m_e e^4}{16\pi^2 \varepsilon_0^2 \hbar^2}$	≈ 27.212 eV	$= 1\,\frac{\hbar^2}{a_0^2\,m_e}$
$\mathcal{E}_{\mathrm{at}} = \frac{e\,e_r}{4\pi\varepsilon_0 a_0^2}$	$\approx 5.1427 \times 10^{11}\,\mathrm{V\,m}^{-1}\,e_r$	$= 1\,\frac{\hbar^2}{a_0^3 m_e e}e_r$
$I_{\mathrm{at}} = \frac{c\varepsilon_0 \mathcal{E}_{\mathrm{at}}^2}{2}$	$\approx 3.5101 \times 10^{20}\,\mathrm{W\,m}^{-2}$	$\approx 5.4556\,\frac{\hbar^3}{a_0^6 m_e^2}$
$t_0 = \frac{a_0}{v_0}$	$\approx 2.4189 \times 10^{-17}$ s	$= 1\,\frac{m_e a_0^2}{\hbar}$

An overview of the units mentioned can be found in Table 4.1. As an example, we use this table to convert frequency into wave length. For frequencies given in a.u. ($\omega = X$ a.u.), the wave length (in SI units) is

$$\lambda = \frac{2\pi c}{\omega} \approx \frac{2\pi\,137.036}{X}\ \text{a.u.} \approx \frac{45.5636}{X}\ \text{nm}$$

Notes and Further Reading

The theory of atoms (and also molecules), including their interaction with laser fields, is covered in the textbook by Bransden and Joachain [37]. The second edition of this book contains a chapter on the interaction with strong fields, which covers some topics that we have omitted. This is the cooling of atoms with lasers and nondipole and relativistic effects. Furthermore, we have not considered linear and nonlinear optical spectroscopy here. They are dealt with in detail in the textbooks by Weissbluth [1] and Mukamel [38].

A Floquet theoretical study of the interaction of an atom with an intense laser pulse is given in [39]. Furthermore, chaos in atomic systems, which can be induced by external laser fields, is covered in the books by Blümel and Reinhard [6], by Reichl [40], and by Bayfield [41].

References

1. M. Weissbluth, *Photon–Atom Interactions* (Academic, New York, 1989)
2. E. Merzbacher, *Quantum Mechanics*, 2nd edn. (Wiley, New York, 1970)
3. I.S. Gradshteyn, I.M. Ryzhik, *Table of Integrals Series and Products*, 5th edn. (Academic, New York, 1994)
4. R. Loudon, Am. J. Phys. **27**, 649 (1959)
5. U. Schwengelbeck, F.H.M. Faisal, Phys. Rev. A **50**, 632 (1994)
6. R. Blümel, W.P. Reinhardt, *Chaos in Atomic Physics* (Cambridge University Press, Cambridge, 1997)
7. J.H. Eberly, Phys. Rev. A **42**, 5750 (1990)
8. F.A. Ilkov, J.E. Decker, S.L. Chin, J. Phys. B: At. Mol. Opt. Phys. **25**, 4005 (1992)
9. L.D. Landau, E.M. Lifshitz, *Lehrbuch der Theoretischen Physik*, vol. III, Quantenmechanik, 8th edn. (Akademie-Verlag, Berlin, 1988)
10. M.V. Ammosov, N.B. Delone, V.P. Krainov, Sov. Phys. JETP **64**, 1191 (1986)
11. L.V. Keldysh, Sov. Phys. JETP **20**, 1307 (1965)
12. P. Lambropoulos, X. Tang, Adv. At. Mol. Phys. Suppl. **1**, 335 (1992)
13. G. van de Sand, J.M. Rost, Phys. Rev. A **62**, 053403 (2000)
14. G. van de Sand, *Semiklassische Dynamik in starken Laserfeldern*, PhD Thesis (Universität Freiburg, 1999)
15. P. Agostini, F. Fabre, G. Mainfray, G. Petite, N.K. Rahman, Phys. Rev. Lett. **42**, 1127 (1979)
16. H.R. Reiss, Opt. Express **8**, 99 (2001)
17. R. Grobe, C.K. Law, Phys. Rev. A **44**, R4114 (1991)
18. R.R. Jones, D. You, P.H. Bucksbaum, Phys. Rev. Lett. **70**, 1236 (1993)
19. D.S. Citrin, Opt. Express **1**, 376 (1997)
20. M.T. Frey, F.B. Dunning, C.O. Reinhold, S. Yoshida, J. Burgdörfer, Phys. Rev. A **59**, 1434 (1999)
21. D. Dimitrovski, E.A. Solov'ev, J.S. Briggs, Phys. Rev. A **72**, 043411 (2005)
22. R. Gebarowski, J. Phys. B: At. Mol. Opt. Phys. **30**, 2143 (1997)
23. C.L. Stokely, J.C. Lancaster, F.B. Dunning, D.G. Arbó, C.O. Reinhold, J. Burgdörfer, Phys. Rev. A **67**, 013403 (2003)
24. L.D. Landau, E.M. Lifschitz, *Lehrbuch der Theoretischen Physik*, vol. I, Mechanik, 10th edn. (Akademie-Verlag, Berlin, 1981)
25. S. Yoshida, F. Grossmann, E. Persson, J. Burgdörfer, Phys. Rev. A **69**, 043410 (2004)
26. S. Yoshida, *Quantum chaos in the periodically kicked Rydberg atom*, PhD Thesis (University of Tennessee, Knoxville, 1999)
27. N.T. Maitra, J. Chem. Phys. **112**, 531 (2000)
28. M. Protopapas, C.H. Keitel, P.L. Knight, Rep. Prog. Phys. **60**, 389 (1997) and references therein
29. J. Seres, E. Seres, A.J. Verhoef, G. Tempea, C. Streli, P. Wobrauschek, V. Yakovlev, A. Scrinzi, C. Spielmann, F. Krausz, Nature **433**, 596 (2005)
30. P.B. Corkum, Phys. Rev. Lett. **71**, 1994 (1993)
31. K. Burnett, V.C. Reed, J. Cooper, P.L. Knight, Phys. Rev. A **45**, 3357 (1992)
32. M. Protopapas, D.G. Lappas, C.H. Keitel, P.L. Knight, Phys. Rev. A **53**, R2933 (1996)
33. V. Yakovlev, A. Scrinzi, Phys. Rev. Lett. **91**, 153901 (2003)

34. T. Kreibich, M. Lein, V. Engel, E.K.U. Gross, Phys. Rev. Lett. **87**, 103901 (2001)
35. G. van de Sand, J.M. Rost, Phys. Rev. Lett. **83**, 524 (1999)
36. R. Abrines, I.C. Percival, Proc. Phys. Soc. **88**, 861 (1966)
37. B.H. Bransden, C.J. Joachain, *Physics of Atoms and Molecules*, 2nd edn. (Pearson Education, Harlow, 2003)
38. S. Mukamel, *Principles of Nonlinear Optical Spectroscopy* (Oxford University Press, New York, 1995)
39. T. Millack, V. Veniard, J. Henkel, Phys. Lett. A **176**, 433 (1993)
40. L.E. Reichl, *The Transition to Chaos*, 2nd edn. (Springer, Berlin Heidelberg New York, 2004)
41. J.E. Bayfield, *Quantum Evolution, An Introduction to Time-Dependent Quantum Mechanics* (Wiley, New York, 1999)

5

Molecules in Strong Laser Fields

Molecules, compared to atoms, have additional degrees of freedom. Not only do the electrons move around the nuclei, but also the nuclei move relative to each other. This allows for a multitude of new phenomena, already without the action of external fields.

By coupling an external electric field to the molecular dynamics, such diverse topics as femtosecond spectroscopy, control of molecular dynamics, and the realization of quantum logic operations emerge. Before entering that wide field, some basics of molecular theory will be repeated by discussing the simplest molecule, the hydrogen molecular ion.

5.1 The Molecular Ion H_2^+

To understand laser driven molecular dynamics, we first review the basics of molecular theory. The time-independent Schrödinger equation of the hydrogen molecular ion is approximately as well as exactly [1] solvable. To keep the discussion simple, we first review the approximate treatment of H_2^+ in the stationary case, leading to the notion of electronic potential energy surfaces, which will be central for the understanding of almost all the material presented in this chapter.

5.1.1 Electronic Potential Energy Surfaces

The hydrogen molecular ion consists of two protons, apart by a distance of R, and a single electron with the distances r_a and r_b to proton a and proton b, respectively. A schematic representation of the molecule is given in Fig. 5.1.

In the following, we try to understand how the energy of the electron is changing as a function of the internuclear distance. It will turn out that there are several possible solutions to this problem, and the outcome has a very intuitive meaning. To make progress, the electron shall first be close to

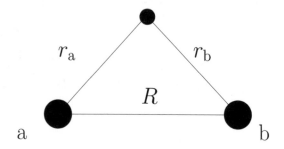

Fig. 5.1. H_2^+ molecular ion consisting of two protons a and b, with distance R, and a single electron with the distances r_a and r_b from the protons

either nucleus a, or nucleus b, while keeping the two nuclei very far apart, i.e. $R \to \infty$. These two limiting cases are simple hydrogen atoms, undisturbed by the other proton, for which we know the exact solution of the respective time-independent Schrödinger equation

$$\hat{H}_a \psi_a = E_a \psi_a \tag{5.1}$$
$$\hat{H}_b \psi_b = E_b \psi_b. \tag{5.2}$$

Furthermore, the two hydrogen atoms are completely equivalent, and therefore the energies are degenerate $E_a = E_b$.

The *full* electronic eigenvalue problem, i. e. the time-independent Schrödinger equation for H_2^+ with a fixed (finite) distance R is given in atomic units by

$$\left\{ -\frac{1}{2}\Delta - \frac{1}{r_a} - \frac{1}{r_b} \right\} \psi_{el}(r_a, r_b, R) = E(R)\psi_{el}(r_a, r_b, R), \tag{5.3}$$

where the Laplacian can be expressed either by $\nabla_{r_a}^2$ or by $\nabla_{r_b}^2$. The internuclear repulsion leads to an R-dependent shift of the energy scale and shall be neglected for the time being. Because we know the solution of the problem for $R \to \infty$, let us calculate the energy $E(R)$ and the corresponding eigenfunction ψ_{el}, as a linear combination of atomic orbitals (LCAO) of the two hydrogen atoms. In the simplest case we can use just a single hydrogen 1s ground state function from Sect. 4.1

$$\psi_{a,b} = \frac{1}{\sqrt{\pi}} \exp\{-r_{a,b}\} \tag{5.4}$$

per proton. The LCAO Ansatz is then given by

$$\psi_{el} = c_1 \psi_a + c_2 \psi_b. \tag{5.5}$$

Inserting it into the time-independent Schrödinger equation yields

$$\left(\hat{H}_a - \frac{1}{r_b} \right) c_1 \psi_a + \left(\hat{H}_b - \frac{1}{r_a} \right) c_2 \psi_b = E(c_1 \psi_a + c_2 \psi_b). \tag{5.6}$$

The application of the Hamiltonian to the ground state leads to a multiplication of the wavefunction by the ground-state energy E_0, and therefore the equation above can be rewritten as

$$\left(E_0 - E - \frac{1}{r_b}\right) c_1\psi_a + \left(E_0 - E - \frac{1}{r_a}\right) c_2\psi_b = 0. \tag{5.7}$$

This equation can be transformed into a linear system of equations in the usual way by multiplying it from the left with the real eigenfunctions $\psi_{a,b}$ and integration over space. The following definitions are appropriate

1. Overlap integral:
   $\int dV \psi_a \psi_b =: S(R) = (1 + R + R^2/3)\exp\{-R\}$
2. Coulomb integral:
   $-\int dV \psi_a \frac{1}{r_b}\psi_a =: C(R) = -(1 - (1 + R)\exp\{-2R\})/R$
3. Exchange integral (having no classical analog):
   $-\int dV \psi_a \frac{1}{r_a}\psi_b =: D(R) = -(1 + R)\exp\{-R\}$
4. Energy difference:
   $\Delta E(R) = E_0 - E(R)$

Exercise 5.1 *Calculate the integrals needed for the LCAO solution procedure for the electronic eigenvalue problem of H_2^+ with the help of the hydrogen 1s functions. Use ellipsoidal coordinates $\mu = (r_a + r_b)/R, \nu = (r_a - r_b)/R, \varphi$, for which the volume element is given by $dV = \frac{1}{8}R^3(\mu^2 - \nu^2)d\mu d\nu d\varphi$, and where $1 \le \mu \le \infty, -1 \le \nu \le 1, 0 \le \varphi \le 2\pi$.*

The coefficients then fulfill the linear system of equations

$$[\Delta E(R) + C(R)]\, c_1 + [\Delta E(R)S(R) + D(R)]\, c_2 = 0 \tag{5.8}$$
$$[\Delta E(R)S(R) + D(R)]\, c_1 + [\Delta E(R) + C(R)]\, c_2 = 0. \tag{5.9}$$

The eigenvalue problem defined above is a generalized one, due to the fact that the eigenvalues $E(R)$ are also appearing in the off-diagonals because of the nonvanishing overlap integral. The condition for solubility leads to the symmetric so-called $1\sigma_g$ solution

$$c_1 = c_2 = \left(\frac{1}{2(1 + S(R))}\right)^{1/2} \tag{5.10}$$

and the antisymmetric $1\sigma_u$ solution

$$c_1 = -c_2 = \left(\frac{1}{2(1 - S(R))}\right)^{1/2} \tag{5.11}$$

with the corresponding energies $(E_0 = 0)$

$$E_\pm(R) = \frac{C(R) \pm D(R)}{1 \pm S(R)}. \tag{5.12}$$

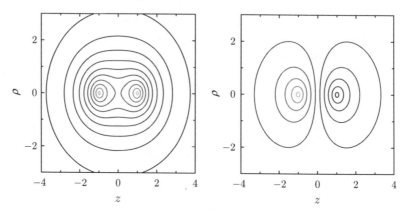

Fig. 5.2. (*Left panel*) symmetric (binding) and (*right panel*) antisymmetric (antibinding) LCAO solution for nuclei located at $R_{a,b} = \pm 1$ as a function of cylindrical coordinates (ρ, z), with the z-axis along the nuclear axis

The eigenfunctions displayed in Fig. 5.2 have decisively distinct character. In the symmetric superposition a sizable part of the wavefunction is located between the two nuclei, whereas the antisymmetric solution $1\sigma_u$ with the minus sign has a node between the nuclei!

For the energies, it is important to note that $C(R)$ as well as $D(R)$ are negative, and therefore the energy is reduced in the symmetric case when compared with two infinitely separated nuclei. To discuss the binding character of the solutions, we have to include the nuclear repulsion in our discussion, however, by considering the quantity

$$E_{\text{tot}}(R) = \frac{C(R) \pm D(R)}{1 \pm S(R)} + \frac{1}{R}. \tag{5.13}$$

For $R \to 0$ the nuclear repulsion dominates due to the fact that both C and D are finite in that limit, see Fig. 5.3. Furthermore, as it should be, for $R \to \infty$ both curves have the H+p case with energy $E_0 = 0$ as the limiting case. At an intermediate value of R_e, the symmetric solution displays a minimum in the energy curve, with a binding energy (dissociation threshold) of D_e, whereas the antisymmetric one is a continuously decreasing function of R as can be seen in Fig. 5.3. A comparison of experimental and theoretical values for the two parameters of the binding potential can be found in Table 5.1.

We have reviewed an "electronic structure calculation" for a molecule with only a single electron, and furthermore, we have used the smallest possible set of basis functions for this system. Nevertheless, the results are quite reasonable as could be seen by comparing them with experimental results. In general, many electron systems have to be considered, however. A prototypical diatomic example is the hydrogen molecule with two electrons. The methods that are typically used to obtain the electronic wavefunction of that molecule are the molecular orbital method or the Heitler-London method. A discussion

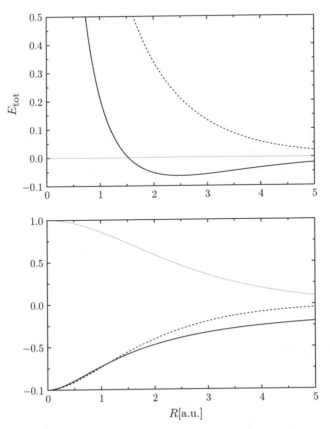

Fig. 5.3. *Upper panel*: LCAO binding (*solid line*) and antibinding (*dashed line*) electronic potential energy curve of H_2^+ (in atomic units) as a function of internuclear distance in atomic units; please note that at the minimum of the binding curve (around 2.5 a.u.), the binding effect is much smaller than the antibinding effect. The zero of energy is indicated by a *dotted line* and corresponds to the H+p case. *Lower panel*: Coulomb integral (*solid line*), exchange integral (*dashed*), and overlap (*dotted*)

Table 5.1. Comparison of experimental and theoretical LCAO results for the equilibrium distance and the dissociation energy of H_2^+

	Theory (LCAO)	Experiment
R_e	130 pm	106 pm
D_e	1.75 eV	2.79 eV

of both approaches can, e.g., be found in [2]. Going beyond the treatment of simple, small systems is done in the field of quantum chemistry, which is dealing with the calculation of the electronic energy curves as a function of internuclear distances in the general case [3].

For our present case, a method to dramatically improve the results by using a slightly modified basis set shall be mentioned. Finkelstein and Horowitz [4] have shown that the variation of the 1s basis functions according to

$$\psi_{a,b} \sim \exp\{-\alpha(R)r_{a,b}\}, \qquad (5.14)$$

where $\alpha(R)$ is allowed to vary between the helium value of 2 at $R = 0$ and the hydrogen value of 1 at $R = \infty$ does improve the results tremendously to $R_e = 106\,\mathrm{pm}$ and $D_e = 2.35\,\mathrm{eV}$. This is a version of the so-called variational LCAO method. An alternative would be to use many more basis function in the standard LCAO method. The convergence of this approach to the experimental value is rather slow, however.

5.1.2 The Morse Potential

Another, arguably more crude, possibility to construct an analytic binding potential energy curve for a diatomic molecule (the two nuclei shall have the masses M_a and M_b) is to use the prototypical Morse potential [5]

$$V_M(R) \equiv D_e[1 - \exp\{-\alpha(R - R_e)\}]^2, \qquad (5.15)$$

displayed in Fig. 5.4, and determine the free parameters from experimental values. To do this, we use that the kinetic energy of the relative motion is given by

$$T_R = \frac{M_r}{2}\dot{R}^2 \qquad (5.16)$$

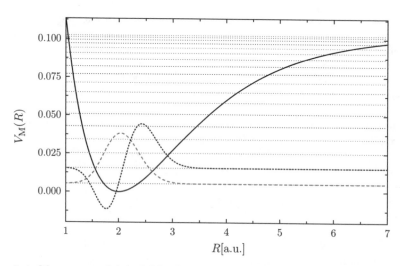

Fig. 5.4. Morse potential (*solid line*) with $D_e \approx 0.103\,\mathrm{a.u.}$ and $R_e \approx 2.00\,\mathrm{a.u.}$ and $\alpha \approx 0.72\,\mathrm{a.u.}$, corresponding to the experimental values of H_2^+ [7], together with the bound eigenvalues and the (unnormalized) two lowest eigenfunctions

with the reduced mass $M_r = M_a M_b/(M_a + M_b)$. Parameters that are available experimentally are e.g.,

- D_0: dissociation energy (from vibrational ground state)
- ω_e: angular frequency of harmonic oscillations around the minimum.

From these, the Morse potential parameters can be extracted according to

- $D_e = D_0 + \hbar\omega_e/2$: dissociation threshold (from minimum of potential curve)
- $\alpha = \omega_e\sqrt{M_r/2D_e}$: range parameter,

where the α parameter follows from the harmonic approximation to the Morse potential.

The solution of the time-independent Schrödinger equation for the Morse potential can be given analytically, although it should be noted that it is only approximate in character [6]. Defining the anharmonicity constant by $x_e = \omega_e/(4D_e)$ the eigenenergies can by calculated according to

$$E_\nu = \omega_e(\nu + 1/2) - x_e\omega_e(\nu + 1/2)^2, \qquad \nu = 0, 1, \ldots. \tag{5.17}$$

Please note that the above definition of the anharmonicity is assuming the exact Morse form of the potential. One could, however, also use the experimental value for x_e, which is typically slightly different from the one that is obtained by inserting the experimental values for ω_e and D_e. In the case of H_2^+, e.g., the frequency and the dissociation energy are given by $\omega_e \approx 0.0105$ a.u. and $D_e \approx 0.103$ a.u. leading to an anharmonicity of $x_e \approx 1/39.2$, whereas the direct experimental value is $x_e \approx 1/37.0$ [7]. Also the range parameter can be calculated from the anharmonicity according to $\alpha = \sqrt{2M_r\omega_e x_e}$. Again α differs slightly if the direct experimental anharmonicity or the one derived from the Morse potential is used!

The bound eigenvalues of the Morse oscillator are depicted in Fig. 5.4 together with the two lowest eigenfunctions according to [6]. Because of the anharmonicity, the distance between the levels decreases with increasing energy (for the parameters of H_2^+ this is barely visible at low quantum numbers). In contrast to the case of the hydrogen atom, where infinitely many levels below the ionization threshold exist, however, only a finite number of levels lies below the threshold of dissociation. The maximal bound state index in the Morse potential can be determined from (5.17) by setting $E_{max} \leq D_e$ and is given by

$$\nu_{max} = \text{Int}(1/(2x_e) - 1/2). \tag{5.18}$$

For the experimental anharmonicity parameter of H_2^+ this number is 18, corresponding to 19 bound states.

5.2 H_2^+ in a Laser Field

The hydrogen molecular ion is the simplest molecule, and therefore it has been the first molecule that has been studied in detail numerically under the influence of an external laser field. Restricting the electronic dynamics to two coupled potential surfaces is allowing the treatment of dissociation via the laser field [8]. Ionization cannot be studied in this framework. We want to focus on both ionization and dissociation phenomena, however. Historically, first the numerical solution of the electron's time-dependent Schrödinger equation with fixed nuclei [9], allowing the investigation of ionization probabilities, has been given. Only after a further increase of computer power, the fully coupled molecular dynamics has been studied [10].

5.2.1 Frozen Nuclei

To study the dynamics of a small molecule under the influence of an external electric field, usually several well-founded assumptions are made. First, one neglects the translational motion of the center of mass and rotations of the molecule. Then the z-axis, along which the nuclei are assumed to be aligned, shall also be the polarization direction of the incident radiation. The effect of molecular alignment [11] is at the heart of this assumption. Fortunately, the problem then has a cylindrical symmetry and the time-dependent Schrödinger equation with fixed nuclei in atomic units is given by

$$i\dot{\Psi}(\rho, z, t) = \left[-\frac{1}{2}\frac{\partial^2}{\partial z^2} + \hat{T}_\rho + V_c(\rho, z) + z\mathcal{E}(t) \right] \Psi(\rho, z, t), \qquad (5.19)$$

where ρ and z are cylindrical coordinates, and the Hamiltonian does not depend on the azimuthal angle φ. Furthermore,

$$\hat{T}_\rho = -\frac{1}{2}\frac{\partial^2}{\partial \rho^2} - \frac{1}{2\rho}\frac{\partial}{\partial \rho} \qquad (5.20)$$

stands for the radial part of the kinetic energy and

$$V_C(\rho, z) = -[\rho^2 + (z - R/2)^2]^{-1/2} - [\rho^2 + (z + R/2)^2]^{-1/2} \qquad (5.21)$$

is the Coulomb potential of the electron binuclear interaction.

The singularity of $\hat{T}_\rho$, as well as of the Coulomb potential at $\rho = 0$, which would lead to numerical difficulties, can be treated very elegantly because of the cylindrical symmetry by expansion in a so-called Fourier-Bessel series [9], see also p. 126 in [12]. If L is the largest distance from the z-axis that is to be described (i.e., the wavefunction shall be zero for $\rho \geq L$), then a complete orthonormal system of functions for the expansion of the ρ-dependence of the wavefunction is given by

$$v_n(\rho) = \frac{\sqrt{2}}{LJ_1(x_n)} J_0(x_n \rho / L), \qquad (5.22)$$

where

$$J_\nu(x) \equiv \left(\frac{x}{2}\right)^\nu \sum_{j=0}^\infty \frac{(-1)^j}{j!\Gamma(j+\nu+1)} \left(\frac{x}{2}\right)^{2j} \tag{5.23}$$

are Bessel functions of νth order and the x_n are the zeros of the Bessel function of 0th order. The basis functions are orthonormalized according to

$$\int_0^L \mathrm{d}\rho\rho v_n(\rho)v_m(\rho) = \delta_{nm}. \tag{5.24}$$

Application of the radial part of the operator of kinetic energy to the basis functions yields

$$\hat{T}_\rho v_n(\rho) = \frac{1}{2}(x_n/L)^2 v_n(\rho), \tag{5.25}$$

i.e. the v_n are eigenfunctions of that operator. This fact can be proven explicitly by using the definition in (5.23).

Exercise 5.2 *Using the definition of the Bessel function of 0th order show that v_n is an eigenfunction of the radial part of the kinetic energy.*

One now expands the wavefunction according to

$$\Psi(\rho,z,t) = \sum_{n=1}^M v_n(\rho)\chi_n(z,t). \tag{5.26}$$

After multiplication of the time-dependent Schrödinger equation from the left with v_k and integration over ρ, the system of partial differential equations

$$\mathrm{i}\dot{\chi}(z,t) = \left[-\frac{1}{2}\frac{\partial^2}{\partial z^2} + \mathbf{A}(z) + z\mathcal{E}(t) \right] \chi(z,t) \tag{5.27}$$

is found for the vector of coefficients and the nonsingular matrix $\mathbf{A}$ (ρV_c is finite at the origin) with the elements

$$A_{kn} = \frac{1}{2}(x_n/L)^2 \delta_{kn} + \int_0^L \mathrm{d}\rho\rho v_k(\rho) V_c(\rho,z) v_n(\rho) \tag{5.28}$$

has been defined.

The time-dependent Schrödinger equation can be solved by again using the split-operator FFT method from Sect. 2.3.2. This leads to the propagated wavefunction vector

$$\chi(z,t+\Delta t) = \exp(-\mathrm{i}\hat{T}_z\Delta t/2)\exp[-\mathrm{i}\mathcal{E}(t')z\Delta t]\exp[-\mathrm{i}\mathbf{A}(z)\Delta t]\exp(-\mathrm{i}\hat{T}_z\Delta t/2)$$
$$\chi(z,t), \tag{5.29}$$

where $\hat{T}_z = -1/2\frac{\partial^2}{\partial z^2}$ and $t' = t + \Delta t/2$. The only difference to what we have already encountered is the fact that the wavefunction is not a scalar but a vector and correspondingly the nondiagonal matrix $\mathbf{A}$ appears in the exponent. To cope with this exponentiated matrix, it is diagonalized using the matrix $\mathbf{U}$ and one finally uses

$$\exp[-i\mathbf{A}(z)\Delta t] = \mathbf{U}\exp[-i\mathbf{A}_D(z)\Delta t]\mathbf{U}^T. \qquad (5.30)$$

The matrix $\mathbf{A}$ is typically of a dimensionality, such that the solution of the eigenvalue problem cannot be done analytically but has to be performed numerically.

Numerical Details and Results

It is worthwhile to note some numerical details of the benchmark calculations of the Bandrauk group [9]. First of all, the numerical grid for the remaining z direction was restricted to $|z| < 128$. Furthermore, it turned out that $L = 8$ and $M = 16$ for the number of Bessel functions was adequate for moderate field intensities around 10^{14}W cm^{-2}. The laser was assumed to be turned on over five cycles of the field with a frequency of $\omega = 0.2\,\text{a.u.}$, corresponding to a wavelength of 228 nm. Finally, the initial state was taken as the electronic ground state $1\sigma_g$. It can be approximated to a good degree by the variational LCAO of the previous section. Alternatively, the LCAO-Ansatz may serve as an initial condition for propagation in imaginary time (see Sect. 2.1.3), distilling the true ground state.

In [9] a measure for ionization was defined with a certain degree of arbitrariness by introducing

$$P_V(t) = 2\pi \int_{-z_I}^{z_I} \mathrm{d}z \int_0^L \mathrm{d}\rho\rho|\Psi(z,\rho,t)|^2 \qquad (5.31)$$

with $z_I = 16\,\text{a.u.}$ as the nonionized part of the probability. Together with the probability to be in the initial state

$$P_0(t) = |\langle\Psi(0)|\Psi(t)\rangle|^2 \qquad (5.32)$$

this quantity is plotted in Fig. 5.5 for two different internuclear distances. In the case $R = 3\,\text{a.u.}$ Rabi oscillations appear for the parameters chosen, which correspond to a one photon transition between the $1\sigma_g$ and $1\sigma_u$ state, whereas for $R = 2\,\text{a.u.}$, two photons would be needed for the same transition. Although we have plotted only the LCAO result for the surfaces in Fig. 5.3, the drastic decrease of the distance between the two surfaces as a function of R can be observed also there.

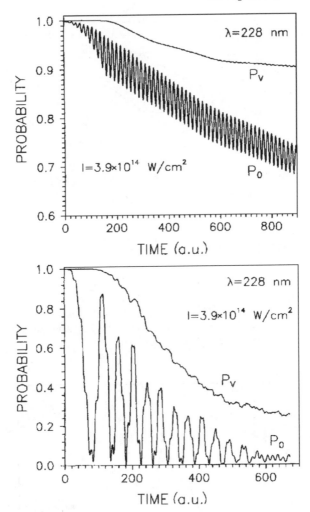

Fig. 5.5. Different probabilities for different fixed internuclear distances as a function of time: *upper panel*: $R = 2$ a.u., *lower panel*: $R = 3$ a.u., from [9]

Charge Resonance Enhanced Ionization

From the quantities defined above, one can extract an ionization rate by using an exponential fit to $P_V(t)$. These rates as a function of intensity can then be compared with the case of the hydrogen atom. In some early work, see e.g., Fig. 5.6, it turned out that the molecular results tend toward the atomic case for increasing excitation, i.e., increasing R from 2 to 3 a.u. [9].

For even larger distances, however, dramatic rate enhancement far beyond the atom limit was found in [13] for a 1,064 nm laser with a five cycle rise. For a fixed intensity this is shown in Fig. 5.7. The explanation of this effect is

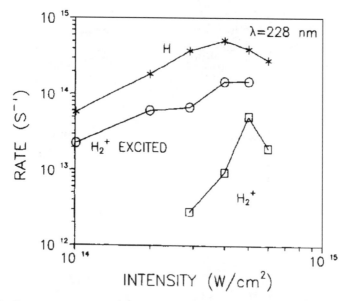

Fig. 5.6. Ionization rate for H_2^+ compared with the atom case for a wavelength of 228 nm as a function of laser intensity. Nuclei fixed at the equilibrium distance (*squares*) and nuclei fixed at $R = 3$ a.u. (*circles*); from [9]

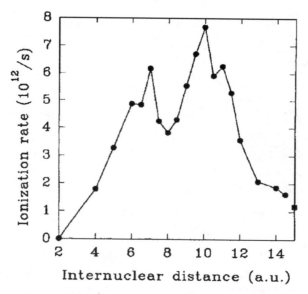

Fig. 5.7. Ionization rate for H_2^+ as a function of internuclear distance for fixed laser intensity of $10^{14}\,\mathrm{W\,cm^{-2}}$ and wavelength of 1,064 nm, atom result indicated by a filled square; from [13]

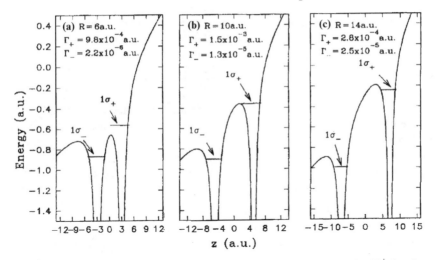

Fig. 5.8. Lowest two static field induced levels and their linewidths for H$_2^+$ for three different internuclear distances: **(a)** $R = 6$ a.u., **(b)** $R = 10$ a.u., **(c)** $R = 14$ a.u.; from [13]

that a pair of charge resonant states (here the almost degenerate $1\sigma_g$ and $1\sigma_u$ states, which have a similar charge distribution at large distances) are strongly coupled to the field at large R, when the dipole moment (see Sect. 5.3.) between them diverges linearly. The effect was therefore termed charge resonance enhanced ionization (CREI).

The success of the static tunneling picture in the atomic case of section 4.2.1 led the authors of [13] to consider the tunneling out of the statically distorted double well potential that the electron experiences in a H$_2^+$ molecule with a field induced potential of $\mathcal{E}_0 z$. The results for the energies evolving out of the two lowest, unperturbed electronic eigenstates are displayed in Fig. 5.8. For the distance $R = 10$ a.u., at which a maximum in the ionization rate can be observed, the width of the upper level has a maximum. In addition, due to the rapid turn on of the field, the population of the upper level is almost equal to the one of the lower level after the amplitude is constant [13] and therefore the system ionizes to a substantial degree.

5.2.2 Nuclei in Motion

Nuclear dynamics in H$_2^+$ can nowadays be treated on the same level as the electronic dynamics by the solution of the full time-dependent Schrödinger equation. The wavefunction then depends on the additional DOF R, describing the relative motion of the nuclei. The coupling to the field shall again be given in the length gauge. In this case, the motion of the center of mass can be separated by introducing the relative coordinate between the two nuclei as well as an electron coordinate, which is measured with respect to the center of mass of the nuclei. This is a lengthily calculation, however, which is reviewed in Appendix 5.A.

As the final result it turns out that the Hamiltonian of equation (5.19) has to be augmented by the kinetic and the potential energy of the nuclei

$$\hat{T}_R + V_R = -\frac{1}{M_p}\frac{\partial^2}{\partial R^2} + \frac{1}{R}, \qquad (5.33)$$

where $M_p/2 \approx 918$ is the reduced nuclear mass in atomic units. Furthermore, as we have seen in Appendix 5.A, the electron mass (which is unity in atomic units) has to be modified slightly to read $m_i = 2M_p/(2M_p + 1)$ and the term with the laser field is to be multiplied by the factor $1/(2M_p + 1)$. Both modifications are marginal due to the large mass ratio, however.

The total wavefunction

$$\Psi(R, \rho, z, t)$$

now also depends on R, and distance dependent quantities like

$$f_1(R, t) = 2\pi \int_0^L d\rho\rho \int_{-z_I}^{z_I} dz |\Psi(R, \rho, z, t)|^2, \qquad (5.34)$$

can be studied. This is the probability density to find the protons a distance R apart and the electrons within a cylinder of height $2z_I$, such that H_2^+ is not fully ionized. z_I is later-on chosen to be 32 a.u. From the R-dependent quantity just defined some integrated quantities can be calculated. These are the dissociation probability, i.e., the probability for the "reaction" $H_2^+ \rightarrow$ H+H$^+$ (the electron could also be evenly distributed between the two protons)

$$P_D(t) = \int_{R_D}^{R_{max}} dR\, f_1(R, t), \qquad (5.35)$$

where again somewhat arbitrarily $R_D = 9.5$ a.u. can be chosen as the onset of dissociation and R_{max} is the nuclear grid boundary. Furthermore, the probability of ionization

$$P_I(t) = 1 - \int_0^{R_{max}} dR\, f_1(R, t), \qquad (5.36)$$

is given by the probability to find the electron outside of a cylinder with $|z| \leq z_I$.

Molecular Stabilization

Numerical results for the case of a molecule initially in the electronic $1\sigma_g$ ground state and in an excited vibronic state with quantum number $\nu = 6$ are depicted in Fig. 5.9. Apart from the probabilities defined above also the probability P_6 to stay in the sixth vibrational state is displayed there. For relatively low intensities (I $= 3.5 \times 10^{13}$ W cm^{-2}) a stabilization of the initial state, i.e., after a short initial decay, an increase of P_6 is observed. It can be

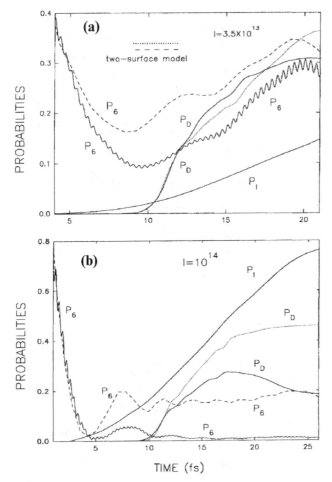

Fig. 5.9. Different probabilities for H$_2^+$ in a laser field ($\lambda = 212$ nm) as a function of time for two different intensities: **(a)** I $= 3.5 \times 10^{13}$ W cm^{-2}, **(b)** I $= 10^{14}$ W cm^{-2}, from [10]

understood due to stimulated emission from the dissociative $1\sigma_u$ state [10]. For higher intensities (I $= 10^{14}$ W cm^{-2}) this effect vanishes, however, because the system is already ionized to a substantial degree. This is in contrast to the prediction of two-state calculations, which are also displayed in Fig. 5.9 and which show the stabilization effect for both intensities. Furthermore, it is worthwhile to note that the fine oscillations in P_6, well visible for low intensities, occur at twice the laser frequency and are due to the counter-rotating term, which is neglected in the RWA (but not in the full numerical calculations reviewed here).

Further light can be shed on the stabilization effect by looking at the nuclear wavefunction at a fixed time. For different field intensities, these results are displayed in Fig. 5.10. Sharp peaks near $R = 3$ and $R = 3.6$ a.u. can

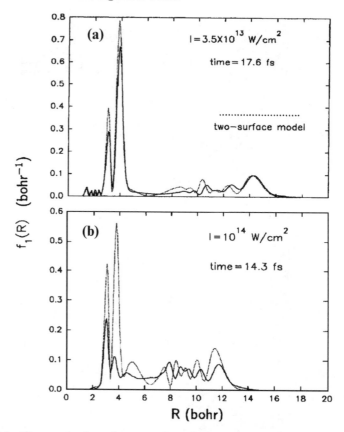

Fig. 5.10. Time-evolved nuclear wavefunction of H_2^+ in a laser field of different intensity (a) I $= 3.5 \times 10^{13}$ W cm^{-2}, (b) I $= 10^{14}$ W cm^{-2}, from [10]. In these plots, also two-surface calculations, which do not account for ionization, are displayed

be observed. In a two state picture the $1\sigma_g$ and $1\sigma_u$ surfaces form avoided crossings[1] if they are dressed by the laser field [14] and the peaks are at the turning points of the bound motion in the upper adiabatic potential. For the lower intensity, the shape of the peaks does not vary much as a function of time, whereas for the higher intensity the peaks decrease considerably due to ionization.

Coulomb Explosion Vs. Dissociation

In the results that we have discussed so far, the initial vibrational state was fixed to be the $\nu = 6$ state. What happens for different initial vibrational quantum numbers?

---

[1] See also Appendix 3.A for avoided crossings of the Floquet analog of dressed states.

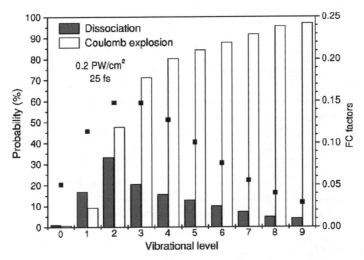

Fig. 5.11. Comparison of the two fragmentation channels as a function of initial vibrational excitation for a 25 fs laser pulse with an intensity of $0.2\,\mathrm{P\,W/cm^2}$ and 800 nm central wavelength, black squares show the overlaps $|\langle H_2^+, \nu | H_2, \nu = 0\rangle|^2$; from [15]

As already discussed, the fragmentation of the molecular ion under the external field can occur via different channels. One channel, in which the electron stays with one (or each) of the two nulcei, is the dissociation channel (without ionization).

$$H_2^+ \rightarrow p + H.$$

The alternative is complete fragmentation, which is our ionization case, given by

$$H_2^+ \rightarrow p + p + e^-.$$

This last channel is also referred to as the Coulomb explosion channel.

The outcome of numerical calculations using a modified 1d soft-core potential, where the softening parameter depends on the internuclear distance [15], and the Crank-Nicholson method has been used for the propagation is displayed in Fig. 5.11. It can be observed that for vibrational levels $\nu \geq 2$ for the field parameters chosen, the Coulomb explosion channel dominates the dissociation channel.

5.3 Adiabatic and Nonadiabatic Nuclear Dynamics

The neutral hydrogen molecule H_2 has an additional electron compared with the hydrogen molecular ion of the previous section. Its dynamics has been treated fully quantum mechanically only recently [16]. To deal with even more complex systems numerically, approximations and/or restrictions of the dynamics to some relevant degrees of freedom have to be made. An important, if not the most important approximation of molecular theory is the Born-Oppenheimer approximation to be discussed in the following.

This approximation is closely related to what we have already done for the case of H_2^+ in Sect. 5.1. There the binding potential energy surface $1\sigma_g$ was taken as a sum of the repulsive potential between the nuclei and the attraction due to the electron in-between the two nuclei. If no other (external) force is acting, then the nuclear motion would be attracted by the minimum of the potential. However, the motion has to be described quantum mechanically, and thus a probability distribution with its maximum at the minimum of the potential curve will result. In the following, we will realize that the notion above is approximate in nature. Even in the case without an external field, the dynamics can, in principle, not be restricted to a single electronic surface.

5.3.1 Born-Oppenheimer Approximation

In (5.3) we have neglected the kinetic as well as the potential energy of the nuclei. The total Hamiltonian for a general molecule with M nuclei and N electrons, however, is given by

$$\hat{H}_{\mathrm{mol}} = \hat{T}_{\mathrm{N}} + \hat{H}_{\mathrm{e}} = \hat{T}_{\mathrm{N}} + \hat{T}_{\mathrm{e}} + V(\boldsymbol{x}, \boldsymbol{X}), \tag{5.37}$$

with the kinetic energies (switching back to SI units for this section)

$$\hat{T}_{\mathrm{N}} = \sum_{i=1}^{M} -\frac{\hbar^2}{2M_i} \frac{\partial^2}{\partial \boldsymbol{R}_i^2} \tag{5.38}$$

$$\hat{T}_{\mathrm{e}} = \sum_{j=1}^{N} -\frac{\hbar^2}{2m_{\mathrm{e}}} \frac{\partial^2}{\partial \boldsymbol{r}_j^2}. \tag{5.39}$$

The potential energy V contains the electron–electron interaction

$$V_{\mathrm{ee}}(\boldsymbol{x}) = \frac{e^2}{4\pi\epsilon_0} \frac{1}{2} \sum_{i \neq j}^{N} \frac{1}{|\boldsymbol{r}_i - \boldsymbol{r}_j|} \tag{5.40}$$

as well as the electron–nucleus (charge Z_i)

$$V_{\mathrm{eN}}(\boldsymbol{x}, \boldsymbol{X}) = -\frac{e^2}{4\pi\epsilon_0} \frac{1}{2} \sum_{j=1}^{N} \sum_{i=1}^{M} \frac{Z_i}{|\boldsymbol{r}_j - \boldsymbol{R}_i|} \tag{5.41}$$

and the internuclear interaction

$$V_{\mathrm{NN}}(\boldsymbol{X}) = \frac{e^2}{4\pi\epsilon_0} \frac{1}{2} \sum_{i \neq j}^{M} \frac{Z_i Z_j}{|\boldsymbol{R}_i - \boldsymbol{R}_j|}. \tag{5.42}$$

All nuclear coordinates are contained in the symbol

$$X = (R_1, \ldots, R_M)$$

whereas the electronic coordinates *including spin* are denoted by

$$x = (r_1 s_1, \ldots, r_N s_N).$$

As usual, the nuclear coordinates are distinguished from the electronic ones by using capital letters for the former and lower case letters for the latter.

The electronic part of the Hamiltonian commutes with the nuclear coordinates, i.e., $[\hat{H}_e, X] = 0$. The nuclear coordinates therefore are "good quantum numbers" for the electronic operator and can be viewed as parameters. One now first solves the electronic eigenvalue problem (the time-independent Schrödinger equation of the electrons), which is the generalized analog of (5.3),

$$\hat{H}_e \phi_\nu = \left[-\sum_j \frac{\hbar^2}{2m_e} \Delta_j + V(x, X) \right] \phi_\nu(x, X) = E_\nu(X)\phi_\nu(x, X), \quad (5.43)$$

where the electronic energy as well as the corresponding wavefunction depend *parametrically* on the nuclear coordinates, and ν is a suitable set of quantum numbers of the electronic system. For the electronic functions orthonormality and completeness relations hold according to

$$\int \mathrm{d}^N x \phi_\nu^*(x, X)\phi_\mu(x, X) = \delta_{\mu\nu} \quad (5.44)$$

$$\sum_\nu |\phi_\nu\rangle\langle\phi_\nu| = 1. \quad (5.45)$$

The *total* wavefunction can thus be expanded by using the electronic states as basis states according to

$$\psi(x, X) = \sum_\nu \phi_\nu(x, X)\chi_\nu(X), \quad (5.46)$$

with the nuclear functions $\chi_\nu(X)$. Inserting this Ansatz into the time-independent Schrödinger equation

$$\left[-\sum_i \frac{\hbar^2}{2M_i} \Delta_i - \sum_j \frac{\hbar^2}{2m_e} \Delta_j + V(x, X) \right] \psi(x, X) = \epsilon \psi(x, X), \quad (5.47)$$

one can now use the electronic Schrödinger equation to "replace" the electronic Hamiltonian and arrives at (written suggestively)

$$\sum_\nu \phi_\nu(x, X) \left[-\sum_i \frac{\hbar^2}{2M_i} \Delta_i + E_\nu(X) \right] \chi_\nu(X) = \sum_\nu \phi_\nu(x, X)\epsilon\chi_\nu(X)$$

$$+ \sum_\nu \sum_i \frac{\hbar^2}{2M_i} [\chi_\nu(X)\Delta_i\phi_\nu(x, X) + 2\nabla_i\chi_\nu(X) \cdot \nabla_i\phi_\nu(x, X)]. \quad (5.48)$$

The second line in the equation above follows from the application of the product rule of differentiation and is due to the change of the electronic states induced by the nuclear motion.

If we neglect this second line and multiply the equation from the left by $\phi_\mu^*(x, X)$, after integration over the electronic coordinates x, we get

$$\left[-\sum_i \frac{\hbar^2}{2M_i} \Delta_i + E_\mu(X) \right] \chi_\mu(X) \approx \epsilon \chi_\mu(X). \tag{5.49}$$

This is the time-independent Schrödinger equation for the nuclear degrees of freedom in a potential that is given by the μth eigensolution of the electrons. We have used this notion already in the last section. As we can appreciate now, an approximation has been made along the way, however. The equation above and also its time-dependent analog, which results in the dynamics on a single potential energy surface are based on the Born-Oppenheimer or adiabatic approximation that resulted due to the neglect of the second line in (5.48)[2].

The approximation that has been made still needs to be justified. That is we have to argue why the terms in the second line of (5.48) might be small compared with the terms in the first line of that equation. Let us just study the terms

$$\frac{\hbar^2}{2M_i} \Delta_i \phi_\nu(x, X)$$

containing the second derivative. The electronic wavefunction depends on X in a similar fashion as it depends on x (depending mainly on the differences $R_i - r_j$). Therefore, the respective derivatives are of the same order of magnitude. The prefactors, however, differ from those in (5.43) by almost 3 orders of magnitude because of the large mass ratio M_i/m_e. That is why the adiabatic approximation is so successful.

In general, however, one has to multiply also the second line in (5.48) from the left with $\phi_\mu^*(x, X)$ and has to integrate over the electronic coordinates. The terms that emerge then lead to transitions between the different electronic surfaces, the so-called nonadiabatic or non Born-Oppenheimer transitions. As we have already stressed, the notion of nuclear motion on a single surface is only approximate and breaks down especially close to avoided crossings of potential surfaces. These transitions can be described by using the Born-Oppenheimer surfaces and allowing for a coupling via the derivative terms of the second line of (5.48). Alternatively, they can also be described in another, so-called diabatic basis. A schematic plot of adiabtic and diabatic levels is given in Fig. 5.12. The diabatic levels can cross and their coupling is given by nondiagonal potential terms and not by derivative terms, which might be advantageous computationally. However, the construction of diabatic surfaces is not unique [18].

---

[2] The original work of Born and Oppenheimer is put into historical context at the end of Chap. 12 in [17].

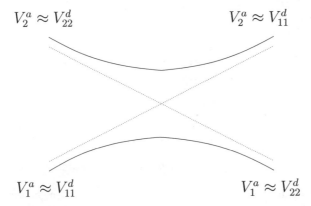

$$V_2^a \approx V_{22}^d \qquad\qquad V_2^a \approx V_{11}^d$$

$$V_1^a \approx V_{11}^d \qquad\qquad V_1^a \approx V_{22}^d$$

Fig. 5.12. Adiabatic (*solid lines*) and diabatic (*dashed lines*) levels in the vicinity of an avoided crossing

Before leaving this section let us come back to the mixed quantum classical methods that we have discussed in Chap. 2. In the framework of an adiabatic description, the classical dynamics is restricted to a single surface, most frequently, the ground state. For frozen nuclear coordinates, the electronic quantum problem is solved for the ground state, and the forces that govern the nuclear motion are determined. After a short time step of the nuclear motion, the same procedure is repeated. In this way, the forces (or the potential surface) are calculated "on the fly". We will not consider this approach and also its generalization to the nonadiabatic case any further but will assume that the potential is at our disposal as the outcome of a quantum chemical calculation.

Relative and Center of Mass Coordinates

Similar to the case of H_2^+, the treatment of the nuclear problem is preferably done in center of mass and relative coordinates. All the relative coordinates shall be contained in the vector $\boldsymbol{R}$. The center of mass is moving freely because the potential does not depend on the corresponding coordinate. We, therefore, concentrate on the description of the relative motion.[3] The relevant masses then are reduced masses. In the diatom case, this has already been used in Sect. 5.2. For more atoms, things become complicated rather quickly. Already in the case of collinear motion of three atoms, cross terms do appear in the kinetic energy.

Exercise 5.3 *For three collinear masses M_1, M_2, M_3 give relative and center of mass coordinates and write the kinetic energy with the help of the canonically conjugate relative momenta.*

---

[3] We will not consider rotational dynamics with the exception of Sect. 5.4.

Coupling to a Laser Field

To couple a laser field to a molecular system, we will use the length gauge. Generalizing the discussion of Sect. 3.1.2 to the many particle case, this leads us to consider the dipole operator

$$d(x, X) = \sum_i Z_i e R_i - \sum_j e r_j. \tag{5.50}$$

After having solved the electronic problem, the dipole matrix element

$$\mu_{\mathrm{ba}}(X) = \int \mathrm{d}^N x \phi_{\mathrm{b}}^*(x, X) d(x, X) \phi_a(x, X) \tag{5.51}$$

has to be calculated. This is a generalization of the dipole matrix element of Sect. 3.2.1, because it still may depend on the internuclear distance.

If we consider different electronic levels, i.e., $b \neq a$, then due to the orthogonality of the corresponding states, only the parts proportional to $\sum_j e r_j$ of the dipole operator survive. If a corresponding transition is not forbidden, then even if one neglects the nonadiabatic terms in the spirit of the adiabatic approximation, still a coupled surface time-dependent Schrödinger equation has to be solved because of the presence of the laser.

5.3.2 Dissociation in a Morse Potential

Before treating the problem of coupled surfaces, the influence of a laser on the wavepacket dynamics in the electronic ground-state shall be studied. Frequently used systems for theoretical calculations are "diatomics" such as HF and CH or OH groups of larger molecules. *Heteronuclear* systems are chosen due to the fact that symmetric homonuclear molecules as e.g., H_2 do not have a dipole moment in the electronic ground-state [2]. If the parameters of a typical infrared laser are chosen appropriately, the diatomic can be driven into dissociation. In contrast to the studies of dissociation of the hydrogen molecular ion of Sect. 5.2, in the following only the nuclear part of the Schrödinger equation will be considered.

After the separation of the center of mass motion, the Hamiltonian for the relative motion of the two nuclei in a Morse potential modeling the electronic ground-state is given by

$$\hat{H} = -\frac{1}{2M_{\mathrm{r}}} \frac{\partial^2}{\partial R^2} + V_{\mathrm{M}}(R) + \mu(R) \mathcal{E}_0 f(t) \cos(\omega(t)t). \tag{5.52}$$

The potential parameters for HF are $D_{\mathrm{e}} = 0.225, R_{\mathrm{e}} = 1.7329, \alpha = 1.1741$ in atomic units [19]. In general, one allows for a chirp, see also Sect. 1.3.3, in the laser frequency.

The R-dependent dipole matrix element (or dipole moment) in principle has to be determined by quantum chemical methods, and we assume it to be given analytically in the form studied in detail by Mecke [20]

$$\mu(R) = \mu_0 R e^{-R/R^*}. \tag{5.53}$$

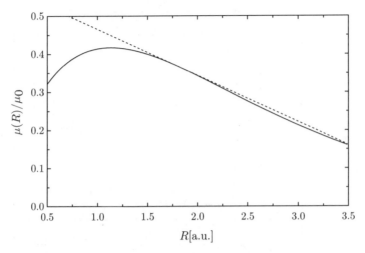

Fig. 5.13. Mecke form for the dipole function (*solid line*) of an X-H diatomic and its linear approximation (*dashed line*) around the minimum of the corresponding Morse potential; parameters used are the ones for the OH stretch in H_2O: $R_e = 1.821$, $R^* = 1.134$ in atomic units [22]

Alternatively, other powers of R than the first may appear in the exponential function. For HF the power of 4 is e.g., frequently used [21]. As can be seen from Fig. 5.13, around the minimum R_e of the Morse potential the dipole function in (5.53) can (up to an irrelevant constant) be approximated by a linear function

$$\mu(R) \approx -\mu'(R - R_e). \tag{5.54}$$

The slope of this linear function is referred to as the dipole gradient or effective charge. For HF this quantity is given by $\mu' = 0.297$ in atomic units [23].

What is the reason for allowing a chirp in the laser frequency? As we have seen in Chap. 3, complete population transfer between two levels is only possible in the case of resonance. So if the oscillator is to be excited resonantly on the ladder of energy levels, in the case of the Morse energies (5.17), the frequency has to decrease as a function of time. Furthermore, the excitation "pulse" has to be a π-pulse. Following these arguments, for the HF molecule, the authors of [23] have constructed an analytic form of a pulse that leads to a large final dissociation probability. The envelope of that pulse and its time varying central frequency are depicted in Fig. 5.14. The frequency decreases from an initial value of ω_{01}[4], equal to the energy difference of the lowest two levels. In the figure also the probabilities for dissociation and the occupation

---

[4] The corresponding vibrational period of HF is 8.4 fs.

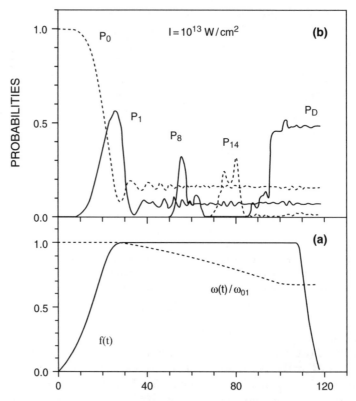

Fig. 5.14. (a) Chirped pulse frequency $\omega(t)$ and envelope function $f(t)$, (b) several probabilities in a driven Morse oscillator, all as a function of time in cycles of the external frequency; adapted from [23]

of different levels of the HF molecule are displayed. The results have been gained by numerically solving the time-dependent Schrödinger equation with the vibronic ground-state as the initial state. Apart from the slight generalization of the chirp, this study is completely analogous to the investigation of multiphoton ionization in the Gauss potential of Sect. 4.2.2.

Analogous results have been found by applying a classical mechanics optimization procedure [24] and are reproduced in Fig. 5.15. It is not surprising that also the classical result displays a down chirp of the frequency. Also in classical mechanics, a softening of the bond occurs for higher energies.

The material just presented is already a glimpse of what we will discuss in detail in Sect. 5.4 on the control of quantum systems. There we will, e.g., review the use of optimal control methods to steer a Morse oscillator into a desired excited vibrational state [25].

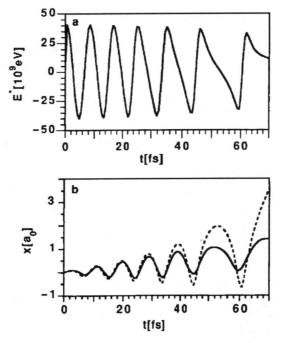

Fig. 5.15. Optimal field (upper panel) and quantum mechanical expectation value of position (*solid line* in lower panel) and classical trajectory (*dashed line* in lower panel) for the dissociation of a Morse oscillator; from [24]

5.3.3 Coupled Potential Surfaces

Now we are ready to deal with the problem of nonadiabatic dynamics on coupled potential energy surfaces. The wavefunction that describes the system in this case is a vector and each component evolves on a specific surface. Because the equations are coupled, the problem is also referred to as a coupled channel problem. In Sect. 5.2, we have encountered such a situation at least formally already. However, the different channels there were the different basis functions in the Fourier-Bessel series expansion.

In this section, we will show how the formalism for the solution of the time-dependent Schrödinger equation has to be augmented to cope with the new situation. First, this will be done fully quantum mechanically and then we will deal in some detail with the semiclassical approximation to coupled surface quantum dynamics.

Quantum Mechanical Approach

As a simple example, let us start with the case of a diatomic molecule and consider two diagonal (diabatic) potential matrix elements $V_{nn}(R), n = 1, 2$.

Their coupling shall be given by arbitrary nondiagonal matrix elements $V_{12}(R,t) = V_{21}(R,t)$, which may depend on time.

The corresponding two surface time-dependent Schrödinger equation is given by[5]

$$i\dot{\chi}_1(R,t) = \left[-\frac{1}{2M_r}\frac{\partial^2}{\partial R^2} + V_{11}(R) \right] \chi_1(R,t) + V_{12}(R,t)\chi_2(R,t), \quad (5.55)$$

$$i\dot{\chi}_2(R,t) = \left[-\frac{1}{2M_r}\frac{\partial^2}{\partial R^2} + V_{22}(R) \right] \chi_2(R,t) + V_{21}(R,t)\chi_1(R,t). \quad (5.56)$$

Its solution can be gained by an extension of the split-operator FFT method of Sect. 2.3. A generalization of the known procedure is necessary due to the fact that the potential is 2×2 matrix

$$\mathbf{V}(R) = \begin{pmatrix} V_{11}(R) & V_{12}(R,t) \\ V_{21}(R,t) & V_{22}(R) \end{pmatrix} \quad (5.57)$$

now, in complete analogy to the matrix $\mathbf{A}$ of Sect. 5.2. To exponentiate it, one first has to diagonalize it, as in the previous section. In contrast to the previous section, in the case of two levels the diagonalization can be done *exactly analytically*, however, leading to [26]

$$\exp\{-i\Delta t\mathbf{V}\} = \exp\left\{ -i\Delta t \left(\frac{V_{11} + V_{22}}{2} \right) \right\} \begin{pmatrix} A & B \\ B & A^* \end{pmatrix}, \quad (5.58)$$

with complex numbers A and B given by

$$A = \cos\phi - i\Delta t\lambda\frac{\sin\phi}{\phi}, \qquad B = i\Delta tV_{12}\frac{\sin\phi}{\phi} \quad (5.59)$$

with the phase

$$\phi(R,t) = \Delta t\sqrt{V_{12}^2(R,t) + \lambda^2(R)} \quad (5.60)$$

and half the energy difference

$$\lambda(R) = \frac{V_{11}(R) - V_{22}(R)}{2}. \quad (5.61)$$

The presented approach is still exact if maximally two surfaces are taking part in the dynamics.

The formalism shall now be applied to the case of coupling between two adiabatic surfaces due to a laser pulse in length gauge using the RWA. The nondiagonal coupling is then given by

$$V_{12}(R,t) = \mu(R)f(t)\mathcal{E}_0\cos(\omega t) \quad (5.62)$$

---

[5] Please note that we are using the symbol for the time-independent nuclear wavefunction also for the time-dependent one.

with the dipole moment μ. To apply the rotating wave approximation of Sect. 3.2.3, the second component of the wavefunction vector is transformed according to

$$\tilde{\chi}_2 = \exp\{i\omega t\}\chi_2. \tag{5.63}$$

Inserting the transformed wavefunction into the time-dependent Schrödinger equation and neglecting the counter-rotating term, the only time-dependence that remains is the one due to the envelope. Furthermore, because of the product rule to be used for the time-derivative, the second surface is shifted by $-\omega$ as displayed in Fig. 5.16. The transformed coupled surface equations in RWA thus contain the modified potential matrix elements

$$\tilde{V}_{22}(R) = V_{22}(R) - \omega \tag{5.64}$$

$$\tilde{V}_{12}(R,t) = \mu(R)f(t)\mathcal{E}_0/2. \tag{5.65}$$

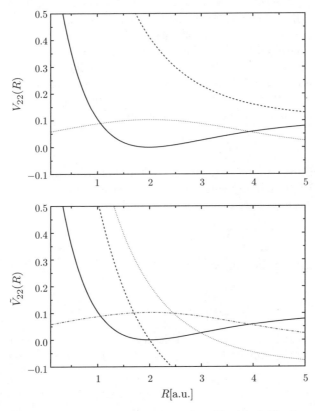

Fig. 5.16. *Upper panel*: Excited electronic surface V_{22} (*dashed*). *Lower panel*: the coupling to a field in the resonance case and for nonresonance leads to differently modified excited states $\tilde{V}_{22}$ (*dashed*: resonance, *dotted*: nonresonance). In both panels also, the ground-state wavefunction (*dashed-dotted*) and the ground-state potential surface V_{11} (*solid*) are depicted

In addition, the Condon approximation can be made. The dipole moment then does not depend on R. Before presenting a first real application of this formalism, we discuss a semiclassical approach to the coupled surface problem.

Semiclassical Approach: Mapping Hamiltonian

The classical and semiclassical description of the motion on coupled potential surfaces seems to be problematic. A way must be found to let trajectories switch from the motion on one to motion on another surface. A method that is ad hoc in nature but is rather successful numerically is the surface hopping technique developed by Tully [27].

Modern semiclassical methods can be *derived* from the underlying quantum mechanics, however, as we will see in the following. These methods have a historical precursor in the so-called "classical electron analog model" by Meier and Miller [28]. We will now review the method by Stock and Thoss, which is based on Schwinger's "mapping formalism" [29]. The N *discrete* electronic levels that shall be taking part in the dynamics are mapped onto N *continuous*, harmonic degrees of freedom in this approach.

The analogy between uncoupled harmonic oscillators and a spin system shall be dealt with in the following for the simple case of $N = 2$. In this case an oscillator of plus type and an oscillator of minus type are defined with the respective annihilation and creation operators

$$\hat{a}_+, \hat{a}_+^\dagger, \qquad \hat{a}_-, \hat{a}_-^\dagger, \tag{5.66}$$

where operators of the same type are fulfilling the standard commutation relations (2.140) and any operators of different type are commuting with each other. Furthermore, occupation number operators

$$\hat{N}_+ = \hat{a}_+^\dagger \hat{a}_+, \quad \hat{N}_- = \hat{a}_-^\dagger \hat{a}_- \tag{5.67}$$

can be defined. Simultaneous eigenkets of $\hat{N}_+$ and $\hat{N}_-$ fulfill the eigenvalue equations

$$\hat{N}_+|n_+, n_-\rangle = n_+|n_+, n_-\rangle, \qquad \hat{N}_-|n_+, n_-\rangle = n_-|n_+, n_-\rangle \tag{5.68}$$

with the eigenvalues $n_\pm$, and an arbitrary state can be created from the vacuum state by the application of $\hat{a}_+^\dagger$ and $\hat{a}_-^\dagger$, using (2.141), according to

$$|n_+, n_-\rangle = \frac{(\hat{a}_+^\dagger)^{n_+}(\hat{a}_-^\dagger)^{n_-}}{\sqrt{n_+!n_-!}}|0, 0\rangle. \tag{5.69}$$

One can now define the products

$$\hat{J}_+ \equiv \hat{a}_+^\dagger \hat{a}_-, \qquad \hat{J}_- \equiv \hat{a}_-^\dagger \hat{a}_+, \qquad \hat{J}_z = \frac{1}{2}(\hat{a}_+^\dagger \hat{a}_+ - \hat{a}_-^\dagger \hat{a}_-), \tag{5.70}$$

where $\hat{J}_z = \frac{1}{2}(\hat{N}_+ - \hat{N}_-)$. It can be shown [30] that these operators fulfill the angular momentum commutation relations

$$[\hat{J}_z, \hat{J}_\pm] = \pm \hat{J}_\pm \tag{5.71}$$

and

$$[\hat{J}_+, \hat{J}_-] = 2\hat{J}_z. \tag{5.72}$$

Furthermore,

$$\hat{\boldsymbol{J}}^2 = \hat{J}_z^2 + \frac{1}{2}(\hat{J}_+\hat{J}_- + \hat{J}_-\hat{J}_+) = \frac{\hat{N}}{2}\left(\frac{\hat{N}}{2} + 1\right) \tag{5.73}$$

with total occupation number operator $\hat{N} \equiv \hat{N}_+ + \hat{N}_-$ holds.

For our purposes the restriction to the subspace $n_+ + n_- = 1$ is appropriate. This is due to the fact that a spin 1/2 particle can be mapped onto two uncoupled oscillators. In case the eigenvalues of the plus and minus oscillator are $n_+ = 1$ and $n_- = 0$, respectively, the particle is in the spin up state. Application of the operator $\hat{J}_-$ leads to $n_+ = 0$ and $n_- = 1$ and the particle is in the spin-down state. The equivalent description of such a system in terms of harmonic oscillators and by using the occupation number representation is depicted in Fig. 5.17.

Now we return to the problem of a Hamilton operator for the coupled dynamics on N surfaces

$$\hat{H} = \sum_{n,m} h_{nm}|\phi_n\rangle\langle\phi_m|. \tag{5.74}$$

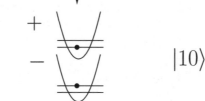

Fig. 5.17. Mapping analogy and occupation number representation

Rewriting this Hamiltonian with the help of continuous bosonic variables, one introduces N harmonic degrees of freedom by using the mapping procedure

$$|\phi_n\rangle\langle\phi_m| \rightarrow \hat{a}_n^\dagger\hat{a}_m, \tag{5.75}$$

$$|\phi_n\rangle \rightarrow |0_1,\ldots 1_n,\ldots,0_N\rangle \tag{5.76}$$

discussed above. This leads to the Hamiltonian

$$\hat{H} = \sum_{n,m} h_{nm}\hat{a}_n^\dagger\hat{a}_m. \tag{5.77}$$

A classical analog of the mapped quantum dynamics can be defined by replacing the position and momentum operators appearing in the creation and annihilation operator

$$\hat{a}_n^\dagger = (\hat{y}_n - \partial/\partial y_n)/\sqrt{2} \tag{5.78}$$

$$\hat{a}_n = (\hat{y}_n + \partial/\partial y_n)/\sqrt{2} \tag{5.79}$$

by the respective classical variables y_n, p_n. The lower case variables $(\boldsymbol{y}, \boldsymbol{p}) = \{y_n, p_n\}$ with $n = 1,\ldots N$ are thus the coordinates and momenta of the auxiliary harmonic oscillators and $(\boldsymbol{R}, \boldsymbol{P})$, are the phase space variables of the relative nuclear motion with the reduced mass M_r. The classical "mapping" Hamiltonian of an N level system is then given by

$$H(\boldsymbol{y},\boldsymbol{p},\boldsymbol{R},\boldsymbol{P}) = \frac{\boldsymbol{P}^2}{2M_r} + H_e, \tag{5.80}$$

with the "electronic" Hamiltonian

$$H_e = \sum_{n=1}^{N} V_{nn}(\boldsymbol{R})\frac{1}{2}(p_n^2 + y_n^2 - 1) + \sum_{n<m=1}^{N} V_{nm}(\boldsymbol{R})(y_n y_m + p_n p_m). \tag{5.81}$$

A semiclassical implementation of the coupled time-dependent Schrödinger equation (5.55,5.56) can now be done by using the Herman-Kluk propagator of Sect. 2.3.4. Each of the vectors $(\boldsymbol{x}, \boldsymbol{p}_i, \boldsymbol{q}_i)$ in the multidimensional formulation of the semiclassical propagator (2.193) contains the nuclear as well as the harmonic degrees of freedom. The initial state is a direct product $|\Psi_\alpha\rangle$ of, e.g., a Gaussian wavepacket in the nuclear coordinate times the ground-state function of the initially unoccupied harmonic mode and the first excited-state in the occupied harmonic mode [29]. The overlap with the coherent state in (2.193) can again be determined analytically.

Application to a Model System

In the following, full quantum as well as semiclassical results of the solution of the coupled surface time-dependent Schrödinger equation (5.55,5.56) will be reviewed for a model that has been used in order to study the breakdown of the Rosen-Zener "approximation" [31]. The dimensionless variables are the same as used there

$$Q \equiv R/R_c \qquad \text{and} \qquad \tau \equiv t/t_c \tag{5.82}$$

with

$$R_c = \sqrt{\hbar/(\sqrt{2}M_r\omega_e)} \qquad \text{and} \qquad t_c = \sqrt{2}/\omega_e. \tag{5.83}$$

Here ω_e is the frequency of the harmonic ground electronic surface and the model potentials are given by

$$V_{11}(Q) = Q^2/2 \tag{5.84}$$

$$\tilde{V}_{22}(Q) = -AQ + B, \tag{5.85}$$

whereas the off-diagonal potential, proportional to the envelope of the external field with dimensionless pulse length parameter T_p, is given by

$$\tilde{V}_{12}(\tau) = D\mathrm{sech}\left[\frac{\tau - \tau_0}{T_p}\right]. \tag{5.86}$$

For an inverse hyperbolic cosine pulse as above, a driven two-level system can be treated analytically and its Rosen-Zener solution has been reviewed in Sect. 3.2.4. The only difference to the case we study here is the absence of the kinetic energy in the Rosen-Zener model. Therefore, although the problem without kinetic energy is solvable exactly analytically, now this solution is an approximation! With these remarks, it is clear that the Rosen-Zener(RZ) approximation is the exact analytical solution of the approximate coupled time-dependent Schrödinger equation

$$i\dot{\chi}_1^{RZ}(\tau) = \lambda(Q)\chi_1^{RZ}(\tau) + \tilde{V}_{12}(\tau)\tilde{\chi}_2^{RZ}(\tau) \tag{5.87}$$

$$i\dot{\tilde{\chi}}_2^{RZ}(\tau) = \tilde{V}_{12}(\tau)\chi_1^{RZ}(\tau) - \lambda(Q)\tilde{\chi}_2^{RZ}(\tau) \tag{5.88}$$

with

$$\lambda(Q) = \frac{\tilde{V}_{22}(Q) - V_{11}(Q)}{2}. \tag{5.89}$$

In the spirit of the so-called Franck-Condon approximation, electronic transitions take place at fixed nuclear positions Q. We can therefore use the solution of Rosen and Zener

$$|\chi_1^{RZ}(\lambda, \tau)|^2 = \left| F\left[DT_p, -DT_p; \frac{1}{2} - i\lambda T_p; z(\tau)\right] \right|^2 \tag{5.90}$$

Table 5.2. Dimensionless model parameters for the problem of two coupled surfaces

	A	B	D	τ_0	T_{p}
Model I	50	10	300	0.1	0.01
Model II	0.1	0.01	2.5	2	0.4

with the hypergeometric function F [32] and

$$z(\tau) = \frac{1}{2}[\tanh(\tau/T_{\mathrm{p}}) + 1]. \tag{5.91}$$

$|\chi_1^{\mathrm{RZ}}|^2$ depends on Q via λ and provides the probability to be in the ground state at a given Q. The total probability to be in the ground state can be gained by multiplying the Rosen-Zener solution with the initial probability density and integrating over position. For the probability to be in the excited state

$$P_2^{\mathrm{RZ}}(\tau) = 1 - \int dQ |\chi_1(Q, -\infty)|^2 |\chi_1^{\mathrm{RZ}}(2\lambda = \Delta V(Q), \tau)|^2. \tag{5.92}$$

follows.

In Table 5.2, the model parameters for the results to be presented are gathered. In both cases, the initial nuclear wavepacket is the ground-state wavefunction of the harmonic surface. The Gaussian part of the 3d wavefunction in the semiclassical case is thus centered around the origin and has the width parameters $(\gamma_{11}, \gamma_{22}, \gamma_{33}) = (2^{-1/2}, 1, 1)$. Quantities of interest are the auto-correlation function

$$c(\tau) = \langle \chi_1(\tau) | \chi_1(0) \rangle + \langle \chi_2(\tau) | \chi_2(0) \rangle \tag{5.93}$$

as well as the occupation probabilities

$$P_{1,2}(\tau) = \langle \chi_{1,2}(\tau) | \chi_{1,2}(\tau) \rangle \tag{5.94}$$

of the different levels.

For model I, a considerable number of Rabi oscillations occurs as can be seen in Fig. 5.18. The quality of the semiclassical results is good. This is so, although the parameters chosen for model I lead to a substantial nonlinearity of the classical equations of motion.[6] Also the semiclassical wavefunctions at time $\tau = 0.135$ in the different levels in Fig. 5.19 give a good account of the quantum wavefunction. Especially, the double humped structure of the wavefunction in the ground state is correctly reproduced.

For model I, the Rosen-Zener approximation (which is not shown) would be very well founded because of the shortness of the pulse, and there would be

---

[6] Both potentials, if uncoupled, would show no nonlinearity in the classical dynamics and would be solvable exactly analytically; when coupled, however, the mapping Hamiltonian is highly anharmonic!

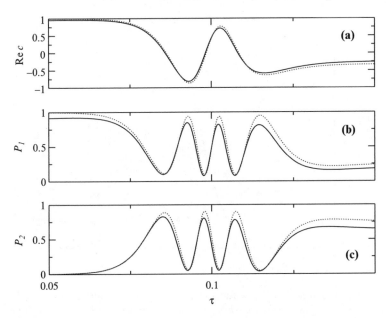

Fig. 5.18. Real part of the autocorrelation function (**a**), and the population of level 1 (**b**), and level 2 (**c**) as a function of time for model I; *solid line*: semiclassical result, *dotted line*: full quantum result; adapted from [33]

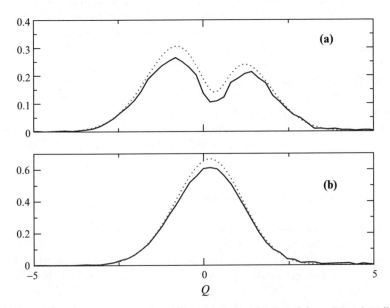

Fig. 5.19. Absolute value of the wavefunction in level 1 (**a**) and level 2 (**b**) at time $\tau = 0.135$ for model I; *solid line*: semiclassical result, *dotted line*: full quantum result; from [33]

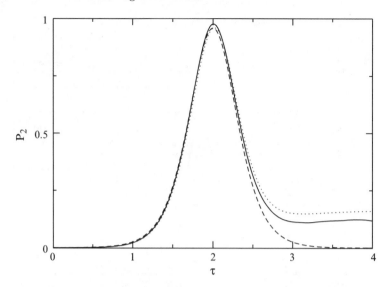

Fig. 5.20. Comparison of the semiclassical (*solid*), the quantum (*long-dashed*) and the Rosen-Zener result (*short dashed*) for the population of level 2 as a function of time for model II, from [33]

almost no difference compared with the full quantum results. Let us consider a case where the neglect of the kinetic energy, sometimes referred to as the short time approximation, breaks down, however. This is the case of model II, which uses the same wavepacket parameters as model I. In Fig. 5.20, a comparison of three different results is displayed. The semiclassical as well as the full quantum and the approximate Rosen-Zener quantum results for the probability to be in the second level are shown. As in case I above, the semiclassical results are representing the full quantum results quite well. In the Rosen-Zener approximation for longer times, a deviation from the quantum result can be observed, however, which is due to the neglect of the kinetic energy in (5.87, 5.88).

5.3.4 Femtosecond Spectroscopy

The study of ultra-fast molecular processes is a rapidly growing research field that has gained considerable attention due to the 1999 Nobel prize in Chemistry for Zewail. Before we concentrate on some theoretical aspects of that field, let us get acquainted with orders of magnitude of time scales in molecules.

To this end, we will convert times into energy (and vice versa) using the formula

$$E/\hbar = 1/t. \tag{5.95}$$

In different units this reads

$$E[\text{eV}] = \frac{0.658}{t[\text{fs}]}, \qquad E[\text{cm}^{-1}] = \frac{5,300}{t[\text{fs}]}. \qquad (5.96)$$

From experimental results, as e.g., molecular spectra, the following ranges for times (periods) in which typical phenomena occur

- Rotation: 10–100 ps
- Normal vibrations: from 100 cm^{-1}, i.e., from 300 fs to 3,000 cm^{-1}, i.e., 10 fs
- Vibrational relaxation: 100 fs–100 ps
- Direct photodissociation: a few 100 fs

can be extracted.

To investigate molecular phenomena on a femtosecond scale, time-resolved measurements, which are frequently referred to as "pump-probe" experiments, are performed. In such experiments, a sample is excited by a first, so-called pump pulse. After a variable time delay T_d, a second, so-called probe pulse is impinging on the excited system and a signal is measured. This may e.g., be the absorption of the system, the fluorescence of the system, or the probability to emit an electron with a certain energy. The time resolution of the experiment is given by the full width at half maximum of the pulses, which shall be used in the following to characterize the shortness of the pulses.

Pump-Probe Spectroscopy of Na$_2$

As a first example of a pump-probe experiment, we consider the excitation of the $(2)^1\Sigma_u^+$-state of the sodium dimer by a 40 fs laser pulse of the fundamental wavelength 340 nm (pump-pulse) and subsequent ionization by a probe-pulse of wavelength 530 nm, arriving after a variable time delay T_d. The energy of the emitted electrons can then be measured as a function of T_d [34]. Theoretical investigations of the same system have been performed by C. Meier in his PhD thesis [35], which we will follow closely.

Before the action of the pump pulse, the molecule is described by the vibronic ground state wavepacket in the electronic ground-state $X^1\Sigma_g^+$, depicted in Fig. 5.21. The perturbation of the system by the pump pulse

$$V_L = \mu f(t)\frac{\mathcal{E}_0}{2}(e^{i\omega_P t} + e^{-i\omega_P t}). \qquad (5.97)$$

leads to excitation of the wavefunction onto an excited electronic surface. In the Condon approximation, the dipole moment μ is assumed to be independent of position. Furthermore, for the following investigation, perturbation theory shall be adequate to describe the laser molecule interaction. In a form applicable to the coupled time-dependent Schrödinger equation, it is reviewed in Appendix 5.B. In first order in the perturbation and in rotating wave approximation (neglecting the counter-rotating term $\sim e^{i\omega_P t}$ of the perturbation)

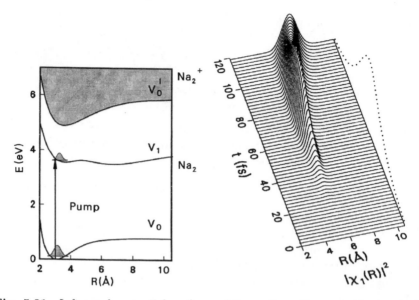

Fig. 5.21. *Left panel*: potential surfaces of the sodium dimer together with the action of the pump pulse onto the initial wavepacket in the electronic ground-state; *right panel*: emergence of the wavepacket on the excited state surface due to the action of the pulse (*dotted line*); adapted from [35]

$$\chi_1(R,t) = \frac{1}{i}\int_0^t dt' e^{-i\hat{H}_1(t-t')}\mu_{10}f(t')\frac{\mathcal{E}_0}{2}e^{-i\omega_P t'}e^{-iE_0 t'}\chi_0(R,0) \quad (5.98)$$

for the wavefunction on the excited state surface is found.

For a numerical implementation, the integral above has to be discretized according to

$$\chi_1(R,t) = \frac{1}{i}\Delta t \sum_{j=0}^{n} e^{-i\hat{H}_1(n-j)\Delta t}\mu_{10}f(j\Delta t)\frac{\mathcal{E}_0}{2}e^{-i(E_0+\omega_P)j\Delta t}\chi_0(R,0). \quad (5.99)$$

The vibronic ground-state with the energy $E = E_0$ is propagated on the shifted ground-state surface until an intermediate time $j\Delta t$ is reached. The resulting state then is propagated on the excited surface with Hamiltonian $\hat{H}_1$ until the final time $n\Delta t$ is reached. It is not known at what time the photon is being absorbed, however, and therefore all the intermediate times have to be integrated over. This procedure can be also formulated iteratively according to [36]

$$\chi_1(R,t_n+\Delta t) = e^{-i\hat{H}_1\Delta t}\chi_1(R,t_n)$$
$$+\frac{\Delta t}{i}\mu_{01}f(t_n+\Delta t)\frac{\mathcal{E}_0}{2}e^{-i(E_0+\omega_P)(t_n+\Delta t)}\chi_0(R,0). \quad (5.100)$$

Here the first term can be calculated with the split-operator method, whereas the second term is given analytically. At each time step, a further part of the wavefunction is lifted on the excited electronic surface. The result of such a calculation is depicted in Fig. 5.21, which shows a vibrationally excited wavepacket on the excited electronic surface that moves almost dispersionless to larger internuclear distances.

Exercise 5.4 *Rewrite the sum in the discretised representation of first order perturbation theory for $n = 1, 2, 3$ and verify the iterative prescription*

$$\chi_1(R, t_n + \Delta t) = \hat{U}_1(\Delta t)\chi_1(R, t_n)$$
$$+ \frac{\Delta t}{i}\mu_{01}\mathcal{E}(t_n + \Delta t)U_0(t_n + \Delta t)\chi_0(R, 0)$$

with $\hat{U}_1(t) = e^{-i\hat{H}_1 t}$, $U_0(t) = e^{-i(E_0 + \omega_P)t}$ and $\mathcal{E}(t) = f(t)\mathcal{E}_0/2$ for the propagation of a wavefunction on two coupled surfaces.

The probe pulse, delayed by a time T_d and centered around the frequency ω_T, now allows the detection of the nuclear wavepacket motion on the excited potential energy surface via the measurement of the energy of the emitted photo electrons after ionization, as can be seen in Fig. 5.22. The key to the understanding of this measurement is the reflection principle. The use of that principle in the theory of photodissociation is reviewed in Appendix 5.C.

To invoke the reflection principle, the wavefunction in the ionization continuum has to be considered. The basis of bound states $\phi_{el,j}(R, r)$ is extended by the continuum states ϕ_{E,V_0^I} (free electron with energy E ionic core in the

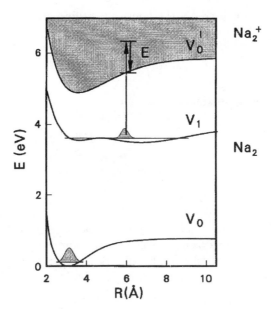

Fig. 5.22. Potential surfaces of the sodium dimer with the action of the probe pulse on the wavepacket in the excited state; from [35]

ground state with potential V_0^I). In the adiabatic approximation, the nuclei would then fulfill the uncoupled equations

$$i\dot\chi_j(R,t) = \{\hat T_R + V_j(R)\}\chi_j(R,t) \tag{5.101}$$

$$i\dot\chi_E(R,t) = \{\hat T_R + V_0^I(R) + E\}\chi_E(R,t). \tag{5.102}$$

Coupling to a laser field in length gauge and using again the RWA we get in first order perturbation theory

$$\chi_E(R,t) = \frac{1}{i\hbar}\int_{T_d}^t dt'\, e^{-i(\hat H_0^I + E - \omega_T)(t-t')/\hbar}\mu_{E1}f(t'-T_d)\frac{\epsilon_0}{2}\chi_1(R,t') \tag{5.103}$$

for the wavefunction in the ionized state. From this, the spectrum of the emitted electron can be extracted according to

$$P^I(E,T_d) = \lim_{t\to\infty}\int dR|\chi_E(R,t)|^2. \tag{5.104}$$

Further progress toward an understanding is made by using the short-time approximation, which we have already encountered in the previous subsection. In this approximation, the kinetic energy of the nuclei is neglected, which means for (5.103):

- Replace $\hat H_0^I$ by V_0^I
- Replace $\chi_1(R,t')$ by $e^{-iV_1(t'-T_d)}\chi_1(R,T_d)$

Equation (5.103) thus becomes the Fourier transformation of the pulse envelope. Using the definition[7]

$$F(x) = \int dt\, e^{ixt}f(t) \tag{5.105}$$

and the short-time approximation, we get

$$P^I(E,T_d) \sim \int dR|\mu_{E1}\chi_1(R,T_d)|^2|F(D(R) + E - \omega_T)|^2. \tag{5.106}$$

Here the definition of the difference potential

$$D(R) = V_0^I(R) - V_1(R) \tag{5.107}$$

has been used. The largest contributions to the expression for the electron emission probability come from regions of vanishing argument of the Fourier transform. This is yet another application of the SPA from Sect. 2.2.1. The SPA condition leads to the definition of so-called transient Franck-Condon regions [37]

---

[7] Please note that the integration boundaries can be shifted to infinity due to the envelope and the terms $e^{iV_1 T_d}$ and $e^{-iV_0^I t}$ drop out by taking the absolute square.

$$D(R_{\mathrm{tr}}) \approx \omega_{\mathrm{T}} - E. \tag{5.108}$$

In case of a monotonous function $D(R)$, the remaining integral can be approximated by

$$P^I(E, T_{\mathrm{d}}) \sim |\mu_{1E}\chi_1(R_{\mathrm{tr}}(E), T_{\mathrm{d}})|^2. \tag{5.109}$$

This is the mathematical formulation of the dynamical reflection principle, saying that the electron spectrum is proportional to the absolute square of the wavefunction at time T_{d}. The structure of its argument tells us that the squared wavefunction is reflected at the difference potential. If it is steep then P^I is broad in energy. If it has a small slope, however, a sharp peak of P^I emerges. Both situations are depicted graphically in Fig. 5.23. In this figure, two different probe pulse delays, $T_1 = 0.2\,\mathrm{ps}$ and $T_2 = 1\,\mathrm{ps}$ are compared with each other. Because of the motion of the wavepacket on surface V_1, a dramatic change of the electron spectrum occurs. For the longer delay, first, it is shifted to smaller energies, and second, it becomes much narrower. By knowing the potential surfaces, one can thus draw conclusions concerning the motion of the nuclear wavepacket from the measured electron spectrum.

The validity of the dynamical reflection principle hinges on the applicability of the short-time approximation. This can be judged by looking at the full time-evolution operators appearing in (5.103)

$$e^{i\hat{H}_0^I t'} e^{-i\hat{H}_1 t'} = e^{\hat{A}}. \tag{5.110}$$

Using the Baker-Campbell-Haussdorf formula from Sect. 2.3.2 in the form

$$e^{\hat{A}} e^{\hat{B}} \approx e^{\hat{A} + \hat{B} + 1/2[\hat{A}, \hat{B}]} \tag{5.111}$$

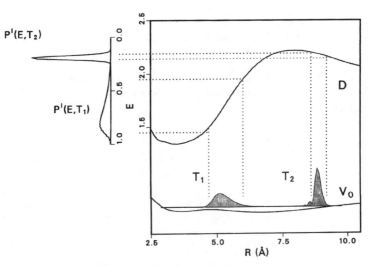

Fig. 5.23. Dynamical reflection principle for two different pulse delays; from [35]

one finds that

$$\hat{\Lambda} = iD(R)t' - [\hat{T}_R, D(R)]t'^2 \qquad (5.112)$$

holds. Retaining only the first term in this expression leads to the short-time approximation. The term proportional to t'^2 with the prefactor

$$[\hat{T}_R, D(R)] \sim \frac{1}{2M_r}(D''(R) + 2D'(R)\partial_R) \sim \frac{i}{M_r}D'p \qquad (5.113)$$

should be small. If one interprets $(p/M_r)t'$ classically as the change of the internuclear distance during the pulse then the condition for the applicability of the short time approximation is

$$D'\frac{p}{M_r}t \ll D(R), \qquad (5.114)$$

i.e., the difference potential should not change much in the range that the wavepacket crosses during the pulse.

Fluorescence Spectroscopy of ICN

Instead of detecting the motion of the wavepacket on the excited surface via the measurement of the energy of emitted electrons as in the previous case, also the fluorescence after excitation into a second excited state can be used. An example that has been studied experimentally as well as theoretically is the ICN molecule. Theoretically, it suffices to consider only the dynamics of the C-I stretch coordinate. The corresponding dynamics on 3 coupled potential surfaces

• Electronic ground state
• I+CN($X^2\Sigma^+$) excited (dissociative) state
• I+CN($B^2\Sigma^+$) excited (dissociative) state

two of which can be seen in Fig. 5.24 has been investigated in [38].
    The coupled surface time-dependent Schrödinger equation for the laser driven system is given by

$$i\hbar\partial_t \begin{pmatrix} \chi_0 \\ \chi_1 \\ \chi_2 \end{pmatrix} = \begin{pmatrix} \hat{H}_0 & \hat{H}_{01} & 0 \\ \hat{H}_{10} & \hat{H}_1 & \hat{H}_{12} \\ 0 & \hat{H}_{21} & \hat{H}_2 \end{pmatrix} \begin{pmatrix} \chi_0 \\ \chi_1 \\ \chi_2 \end{pmatrix}, \qquad (5.115)$$

where in the rotating wave approximation the coupling by the pump, respectively the probe pulse is given by

$$\hat{H}_{01} = \mu_{01}A_1(t)e^{-i\omega_P t} \qquad (5.116)$$

$$\hat{H}_{12} = \mu_{12}A_2(t - T_d)e^{-i\omega_T t}. \qquad (5.117)$$

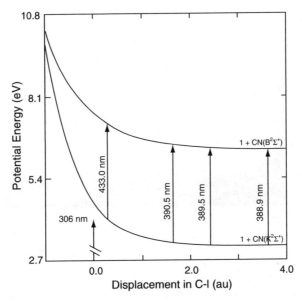

Fig. 5.24. Dissociative potential surfaces and different probe frequencies for ICN, from [38]

The occupation of the second excited-state is proportional to the measured fluorescence signal and is therefore to be calculated theoretically. In first order perturbation theory, we get for the wavefunctions in the different electronic states (if $E_0 = 0$)

$$|\chi_0(t)\rangle = e^{-iE_0 t}|\chi_0(0)\rangle = |\chi_0(0)\rangle \tag{5.118}$$

$$|\chi_1(t)\rangle \sim \int_{-\infty}^{t} dt'\, \mu_{01} A_1(t') e^{-i\hat{H}_1'(t-t')}|\chi_0(0)\rangle \tag{5.119}$$

$$|\chi_2(t)\rangle \sim \int_{-\infty}^{t} dt'\, \mu_{12} A_2(t' - T_d) e^{-i\hat{H}_2'(t-t')}|\chi_1(t)\rangle \tag{5.120}$$

with $\hat{H}_1' = \hat{H}_1 - \omega_P$ and $\hat{H}_2' = \hat{H}_2 - \omega_T$, due to the fact that the potentials are shifted in RWA by ω_P, respectively ω_T. The transfer of probability density to an excited surface can only be large if the crossing with the shifted level is at the maximum of the wavepacket. This so-called resonance case was depicted together with the off-resonance case in Fig. 5.16. Only in the case of resonance a total population transfer is possible by a so-called π-pulse.

In the following the results of a simulation [38] of an experiment of the Zewail group [39] will be reviewed.

- Pump and probe pulse have a FWHM of 125 fs.
- The pump wavelength is fixed at the off-resonant value of 306 nm.
- Four different (only three in the experiment) probe wavelengths have been applied.

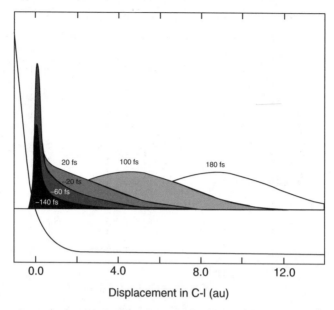

Fig. 5.25. Time evolution on the first excited potential energy surface, from [38]

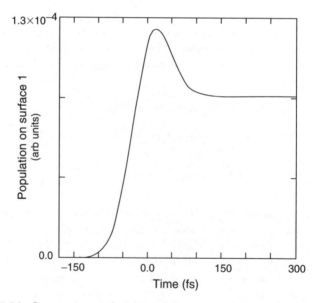

Fig. 5.26. Occupation probability of the first excited surface, from [38]

In Fig. 5.25, the time evolution of the wavepacket on the first excited-state surface is displayed, with the pump pulse being centered around $t = 0$. The wavepacket dissociates and spreads simultaneously. In Fig. 5.26, the population of the first excited-state is depicted. The fact that this population is

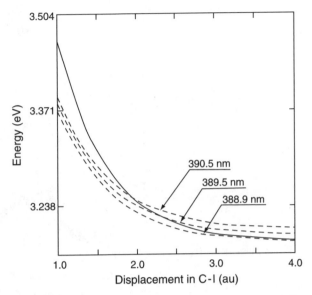

Fig. 5.27. First excited surface and second surface shifted by different amounts, corresponding to different probe frequencies; from [38]

on the order of a few times 10^{-4} reflects the fact that an off-resonant pump frequency is used in the experiment.

The probe part of the experiment has been performed with three different probe frequencies, leading to resonance at different internuclear displacements as depicted in Fig. 5.27. The additional probe wavelength of 433 nm is not depicted in this plot. For the long wavelengths, the resonance condition is more or less well localized in space, and therefore as a function of the probe pulse delay, a peaked structure is to be expected. This is exactly what can be observed in Fig. 5.28! For the additional theoretical wavelength, the signal is barely visible but peaked. The peak tends to become a plateau for the shorter wavelengths. In these cases, the resonance condition is fulfilled for a long interval of internuclear distances as can be seen in Fig. 5.27. The plateau is perfectly developed in the case of 388.9 nm. The experimental results are very well reproduced by the calculations as can be seen by comparing the two panels in Fig. 5.28.

5.4 Control of Molecular Dynamics

Up to now we have encountered a multitude of partly counter-intuitive phenomena, appearing in atomic or molecular systems exposed to a laser field. Quite naturally, one might ask if a suitable field can be found that e.g., drives a system into a desired quantum state or steers a chemical reaction into a desired channel.

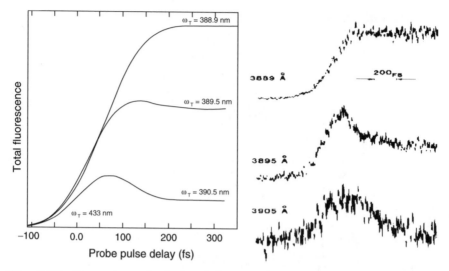

Fig. 5.28. Comparison of calculated (*left panel*) and experimental (*right panel*) fluorescence spectra of ICN for 4 (theory), respectively 3 (experiment) values of the probe wavelength, from [38, 39]

Let us start to find an answer to that question for a system that exhibits one of the most fundamental quantum phenomena: a symmetric double well potential allowing for coherent tunneling between its two minima. This system has been studied in detail under the influence of an external periodic laser field. In the following, we will then refrain from the restriction to periodic fields and will consider pulsed fields with arbitrary pulse shapes that e.g., allow for the control of chemical reactions or for the selective excitation of vibrational modes.

5.4.1 Control of Tunneling

Tunneling in a stationary double well is a phenomenon, which is discussed in almost every textbook on quantum mechanics. In the heyday of quantum theory, it has been, e.g., used to explain the vibrational spectrum of pyramidal molecules, as NH_3, by F. Hund [40]. The influence of a periodic external force on coherent tunneling has been investigated in my PhD thesis, and the results have been published in [41]. The basis for the understanding of those results is Floquet theory as will be seen in the following.

The Model System

A model potential for a particle of mass M_r moving in a symmetric quartic double well is given by

$$V_{DW}(R) \equiv -\frac{M_r \omega_e^2}{4} R^2 + \frac{M_r^2 \omega_e^4}{64 E_B} R^4. \tag{5.121}$$

The frequency of small oscillations around the minima

$$R_{r,l} = \pm \sqrt{\frac{8E_B}{M_r \omega_e^2}} \tag{5.122}$$

of this potential is ω_e, and E_B denotes the height of the barrier between the two wells. The dynamics of a wavepacket, which can be written as a superposition of the ground and the first excited-state of that single surface, is the well-known coherent tunneling dynamics, which is reviewed in Appendix 5.D.

A realization of the potential of (5.121) is given, e.g., by the pyramidal NH_3 molecule. The relevant coordinate R refers to the umbrella mode (see Fig. 1.4 of Chap. 1) and measures the distance between the nitrogen atom and the hydrogen plane. The reduced mass is $M_r = M_N M_{3H}/(M_N + M_{3H})$ (see footnote on p. 566 of [2]). In this system an external periodic force can be generated by a monochromatic laser field of amplitude $\mathcal{E}_0$ coupling to the dipole moment

$$\mu(R) = \mu' R \tag{5.123}$$

of the molecule with the dipole gradient μ'. The amplitude of the external force is then given by

$$F_0 = \mu' \mathcal{E}_0. \tag{5.124}$$

In this subsection, we measure energies in units of $\hbar\omega_e$, such that e.g., $D = E_B/\hbar\omega_e$. Time is measured in units of $1/\omega_e$, $x = \sqrt{(M_r\omega_e)/\hbar}R$, and the dimensionless amplitude is given by $S = F_0/\sqrt{\hbar M_r \omega_e^3}$. The dimensionless Hamiltonian is then given by

$$\hat{H}(x,t) = -\frac{1}{2}\partial_x^2 - \frac{1}{4}x^2 + \frac{1}{64D}x^4 + xS\sin(wt), \tag{5.125}$$

where $w = \omega/\omega_e$ is the dimensionless ratio of driving frequency and harmonic well frequency.

Coherent Destruction of Tunneling

One of the most counter-intuitive effects that external forcing can have is the suppression of the tunneling dynamics in a double well. For frequencies in the middle of the interval

$$\frac{\Delta}{2} \leq w \leq w_{res} \tag{5.126}$$

with $\Delta = E_2 - E_1$ and $w_{res} = E_3 - E_1$ (E_i being the unperturbed energies of the double well), this suppression is found along a one-dimensional curve in the (w, S) parameter space [42].

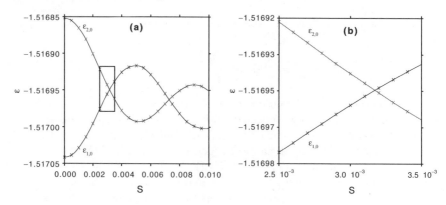

Fig. 5.29. Exact crossing of two quasi-energies of the driven double well ($D = 2, w = 0.01$) (a) $0 < S < 0.01$; (b) zoomed region around $S \approx 3 \times 10^{-3}$, from [42]

The reason for this behavior is an exact crossing of the quasi-energies emerging from the lowest two unperturbed energies as a function of the amplitude. The quasi-energies can be determined by diagonalizing the Floquet matrix of Sect. 2.2.7. For $D = 2$ (which is close to the NH_3 value of 2.18) and an external frequency of $w = 0.01$, near the geometric mean of the unperturbed tunneling frequency $\Delta = 1.895 \times 10^{-4}$ and the first resonance frequency $w_{res} = 0.876$, the behavior shown in Fig. 5.29 is found. In panel (a) of this figure the quasi-energies cross at two different values of the external force. These crossings are exact (see e.g., panel (b)), due to the fact that the quasi-eigenfunctions have different symmetry under the generalized parity transformation defined in (3.69). The non-crossing rule does therefore not hold, and as a function of a parameter the quasi-energies may approach each other arbitrarily closely.

Now the time-evolution of a Gaussian wavepacket, initially centered in the left well, $\chi_l^{GW}(x)$[8] for the parameters at the first exact crossing has been investigated. Quantities of interest are the probability to stay (staying probability) in the initial state, i.e., the absolute square of the auto-correlation function of that wavefunction

$$P(t) := |\langle \chi_l^{GW}(t)|\chi_l^{GW}(0)\rangle|^2 \tag{5.127}$$

and the probability to be to the left of the barrier

$$\rho_l(t) := \int_{-\infty}^{0} dx |\chi_l^{GW}(x,t)|^2. \tag{5.128}$$

These quantities are displayed in Fig. 5.30.

---

[8] Please note that this initial state is slightly different from the one studied in Appendix 5.D.

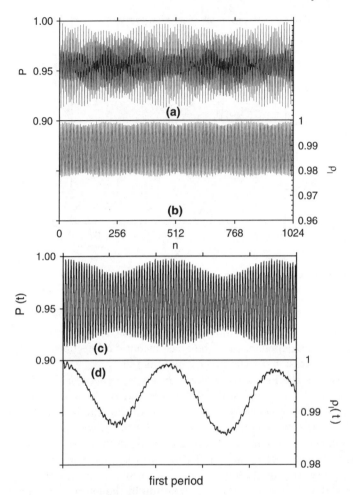

Fig. 5.30. Staying probability and probability to be to the left of the barrier $(D = 2)$ for the exact crossing parameters $S = 3.171 \times 10^{-3}, w = 0.01$: **(a–b)** Stroboscopic time evolution over 2^{10} periods $\frac{2\pi}{w}$; **(c–d)** time evolution inside the first period, adapted from [42]

The unperturbed tunneling period for the parameters chosen here $(D = 2, w = 0.01)$ is at around 50 periods, T, of the external field. In Fig. 5.30, panels (a–b) we can see, however, that in the presence of driving the particle is almost completely localized at the initial position even after $2^{10}T$. This is due to the fact that the initial wavefunction consists mainly of two Floquet functions whose quasi-energies cross exactly. Small deviations of the staying probability from unity are due to the finite overlap of the initial state with other Floquet functions, whose energies are not crossing exactly. The dynamics has been considered only stroboscopically, so far. Time-evolution *during a period of*

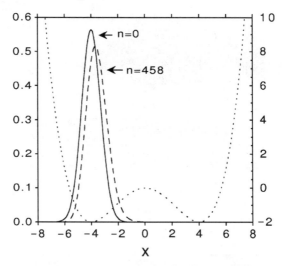

Fig. 5.31. Absolute square of wavefunctions at $t = 0$ (*solid line*) and at $t = 458T$ (*dashed line*), and unperturbed potential (*dotted line*) with $D = 2$, adapted from [42]

the external field is shown in panels (c–d) of Fig. 5.30. It can be observed that the periodic time-dependence of the quasi-eigenfunctions does not destroy the tunneling suppression for the present parameters.

To illustrate the localization effect in position space, in Fig. 5.31, the absolute square of the initial state $\chi_l^{\mathrm{GW}}(x,0)$ and the time-evolved state with the lowest overlap (occurring at $t = 458T$) during the first 1,024 periods is depicted. Apart from a small shift of the center of the wavepacket to the right, there is almost no dynamics observable. This picture also explains why the probability to stay left to the barrier deviates less from unity than the staying probability. In calculating ρ_l, one has to integrate over the whole range $-\infty < x \le 0$ and motion of the wavefunction in the left well will not show up directly in the dynamics of ρ_l. The probability to stay left thus deviates only by maximally 2% from unity, whereas $P(n)$ maximally looses around 8% of its initial value.

Crossing Manifold and Two Level System

As already mentioned, the localization phenomenon occurs along a 1d manifold in the (w, S) parameter space, along which two relevant quasi-energies, having different parity, cross. This manifold has been determined in [42] and is shown in Fig. 5.32. The crossing of the two quasi energies is a necessary but not a sufficient condition for the localization phenomenon, however. This is studied in great detail in [42], where it is shown how the old unperturbed tunneling behavior is recovered for small driving frequencies $w \approx \Delta$ and what happens at resonance $w \approx w_{\mathrm{res}}$, where the third level comes into play.

A deeper understanding of the linear part of the manifold along which real localization is observed can be gained by studying the two-level system describing just the lowest two levels of the double well problem [43]. The corresponding time-dependent Schrödinger equation is given by

$$i\dot{c}_1(t) = E_1 c_1(t) + \langle \chi_1 | x | \chi_2 \rangle S \sin(wt) c_2(t) \qquad (5.129)$$

$$i\dot{c}_2(t) = E_2 c_2(t) + \langle \chi_1 | x | \chi_2 \rangle S \sin(wt) c_1(t) \qquad (5.130)$$

with $c_{1,2}(t) \equiv \langle \chi_{1,2} | \chi(t) \rangle$. The unperturbed Hamiltonian is now a 2×2 matrix and the quasi-energies can again be determined according to the scheme reviewed in Sect. 2.2.7. The location in parameter space of the first exact crossing of the two quasi-energies, emerging out of the two lowest unperturbed states, is plotted in Fig. 5.32 as a solid line. For frequencies $w \ll w_{\mathrm{res}}$ it is very close to the manifold of the full problem, and it stays a perfect straight line also for frequencies $w > w_{\mathrm{res}}$ due to the nonexistence of a third unperturbed level.

The slope of the manifold in the linear range can be determined analytically for a two-level system. In the case of $w \gg \Delta$ (defining the linear region), it has been shown by Shirley that the first crossing is approximately given by the first zero of the Bessel function $J_0(\frac{2b}{w})$ [44]. Here b denotes the field strength multiplied by the dipole matrix element

$$b \equiv \langle \chi_1 | x | \chi_2 \rangle S \overset{D=2}{\approx} 3.791 S. \qquad (5.131)$$

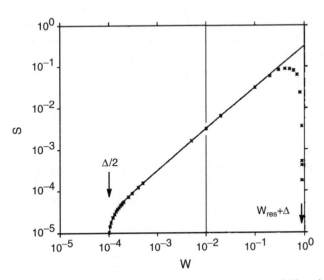

Fig. 5.32. Double logarithmic plot of the one-dimensional manifold in (w, S) parameter space along which the relevant quasi-energies cross for the first time. Driven double well $(D = 2)$: *crosses*; driven two-level system: *solid line*. The *vertical line* crosses the manifold at $w = 0.01, S = 3.171 \times 10^{-3}$, from [42]

Using the simple form of the argument of the Bessel function and its first zero [45] the straight line

$$S = \frac{2.40482\ldots}{2\langle\chi_1|x|\chi_2\rangle}w \overset{D=2}{\approx} 0.3172w \tag{5.132}$$

in (w, S) parameter space is found. Higher zeroes of the Bessel function give straight lines along which the quasi-energies cross each other exactly again (see e.g. Fig. 5.29a). Tuning the parameters to an exact crossing, localization is also found in the time-dependent two-level Schrödinger equation (5.129, 5.130) with the initial conditions $c_1(0) = -c_2(0) = \frac{1}{\sqrt{2}}$.

Furthermore, it is worthwhile to note that in the strong field limit, the suppression phenomenon can also be understood in the transfer matrix formalism [46].

The Asymmetric Double Well Potential

What is the effect of a finite asymmetry on the localization phenomenon? To study this question, a potential of the form

$$V_\sigma(x, t) = -\frac{1}{4}x^2 + \frac{1}{64D}x^4 + \sigma x + xS\sin(wt) \tag{5.133}$$

with a static asymmetry, $\sigma > 0$, can be studied. In this case, no symmetry under the generalized parity transformation (3.69) does exist any more. This leads to the fact that all Floquet energies, due to the non-crossing rule, cannot cross exactly (except maybe at singular points). Therefore no 1d manifold along which two quasi-energies cross does exist.

What happens to the allowed crossings that we have observed in the symmetric case? To answer this question a relatively small asymmetry with parameter σ can be applied. In Fig. 5.33, the same field and potential parameters have been used as in Fig. 5.29b, except for the asymmetry. We can see that the allowed exact crossing becomes an avoided crossing in the presence of asymmetry. Localization therefore goes away gradually. The splitting of the levels is rather small and the wavepacket would still be localized for relatively long times. It would not be localized forever any more as in the symmetric case, however. For strong asymmetry two effects have to be considered. First, the avoided crossing becomes broader, but second, a partial localization does result from the fact that the lowest eigenstate becomes similar to the coherent state in the lower well.

More Driven Double Well Systems

The realization of driven double well systems is possible in many different branches of physics. Recently, driven double wells have e.g., been realized in

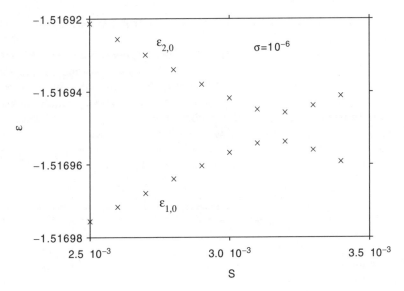

Fig. 5.33. Quasi-energies as a function of S for $D = 2$ and $w = 0.01$ in the asymmetric $(\sigma = 10^{-6})$ double well; from [42]

optical fiber systems. In such a system, a light beam propagating through a periodically curved waveguide is coupled to a parallel fiber. In this setup, the first experimental realization of the effect of coherent destruction of tunneling has been performed [47]!

Another physical system whose dynamics can be described with the help of a multistable potential is the rf-SQUID. There the macroscopic flux through the ring is the tunneling degree of freedom. An external perturbation maybe given by a magnetic field. However, this is an example from the realm of solid-state physics, and shall not be dealt with here. Very recently, the direct observation of suppression of single particle tunneling of atoms in light shift double well potentials has been reported [48].

Let us finally come back to molecular physics. Apart from the NH_3-molecule where the nitrogen atom experiences a double well potential, also the electron in the H_2^+- (resp. D_2^+-) molecule sees a double well potential, because of the electron–nuclear interaction. Recently, it has been shown experimentally that the dissociation of the electron can be steered by the carrier envelope phase such that the electron is localized preferentially at a specific proton (deuteron) [49, 50].

5.4.2 Control of Population Transfer

The transfer of population into a desired state is one of the central challenges of control theory. Before we discuss a direct approach to that field, using optimal control theory, a counter-intuitive method to control population transfer shall be reviewed.

In molecular systems, this is the stimulated Raman adiabatic passage or short STIRAP method. In this scheme a three level system, displayed in Fig. 5.34, is coupled via two different laser pulses. A direct coupling of level $|1\rangle$, which might by a rotational level in the vibrational ground state and the highly excited vibrational state $|3\rangle$ shall e.g., be dipole forbidden. The methodology is used experimentally to selectively excite vibrational states [51].

The pump-pulse couples levels 1 and 2, while the Stokes pulse couples levels 2 and 3. The total Hamilton matrix $\mathbf{H} = \mathbf{H_0} + \mathbf{W}$ is given by

$$\mathbf{H} = \begin{pmatrix} E_1 & -\mu_{12}E_P\cos(\omega_P t) & 0 \\ -\mu_{21}E_P\cos(\omega_P t) & E_2 & -\mu_{23}E_S\cos(\omega_S t) \\ 0 & -\mu_{32}E_S\cos(\omega_S t) & E_3 \end{pmatrix}. \quad (5.134)$$

Transformation into the interaction picture with the help of the unperturbed Hamiltonian

$$\mathbf{H_0} = \begin{pmatrix} E_1 & 0 & 0 \\ 0 & E_2 & 0 \\ 0 & 0 & E_3 \end{pmatrix}, \quad (5.135)$$

and invoking the rotating wave approximation leads to

$$\mathbf{W_I} = \mathbf{U_0^\dagger} \mathbf{W} \mathbf{U_0} = -\frac{1}{2}\begin{pmatrix} 0 & \Omega_P e^{-i\Delta_P t} & 0 \\ \Omega_P e^{i\Delta_P t} & 0 & \Omega_S e^{i\Delta_S t} \\ 0 & \Omega_S e^{-i\Delta_S t} & 0 \end{pmatrix} \quad (5.136)$$

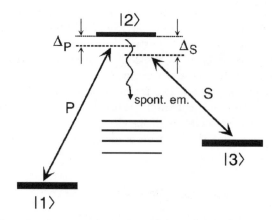

Fig. 5.34. A three level system (so-called Λ-system) coupled via pump and Stokes pulse with the respective detunings Δ_P and Δ_S, from [51]

for the Hamiltonian in the interaction representation. Here the abbreviations $\Omega_P(t) = \mu_{21}E_P(t)$, $\Omega_S(t) = \mu_{32}E_S(t)$ for the Rabi frequencies without detuning and $\Delta_P = (\omega_2 - \omega_1) - \omega_P$, $\Delta_S = (\omega_2 - \omega_3) - \omega_S$ for the detunings that are plotted in Fig. 5.34 have been defined.

In the case of vanishing detunings the time-dependent eigenvalues and eigenstates (dressed states) are given by

$$E_{0,\pm} = 0, \pm \frac{(\Omega_P^2 + \Omega_S^2)^{1/2}}{2} \tag{5.137}$$

$$|g_0\rangle = \cos\Theta|1\rangle - \sin\Theta|3\rangle) \tag{5.138}$$

$$|g_\pm\rangle = \sin\Theta|1\rangle - \cos\Theta|3\rangle) \mp \frac{1}{\sqrt{2}}|2\rangle, \tag{5.139}$$

with $\Omega = (\Omega_P^2 + \Omega_S^2)^{1/2}$ and the definition

$$\Theta \equiv \arctan\left(\frac{\Omega_P}{\Omega_S}\right). \tag{5.140}$$

Cosines and sines of this angle can be resolved according to the relations given in footnote 3 of Chap. 3.

With the help of the dressed states and of the quantum mechanical adiabatic theorem of Appendix 5.E, the pulse sequence can be understood. Starting from state $|1\rangle$ only the dressed state $|g_0\rangle$ is occupied initially if $\Omega_S \gg \Omega_P$. This amounts to the counter-intuitive pulse sequence depicted in panel (a) of Fig. 5.35, where the Stokes pulse comes before the pump pulse! If the field changes adiabatically[9] then according to the adiabatic theorem the system stays in the dressed state $|g_0\rangle$ of the instantaneous Hamiltonian. For large positive times $\Omega_P \gg \Omega_S$ holds, however, and thus the system finally is in state $|3\rangle$, without having occupied the "dark state" $|2\rangle$ in the meantime. This dynamics is depicted in Fig. 5.35, where also the mixing angle and the dressed eigenvalues are displayed.

The ordering of the pulses is counter-intuitive. They have to have a non-vanishing overlap, however, in order for population transfer to be achieved (Fig. 5.35). This can be seen by doing Exercise 15.8 in [17], the reference we are following closely throughout this subsection. Furthermore, an alternative perspective on STIRAP can be gained by demanding constant probability to be in the second (dark) state. For this the time derivative

$$\frac{d|a_2|^2}{dt} = 2\text{Re}[a_2^*\dot{a}_2] = -[\Omega_P(t)\text{Im}(a_2^*a_1) + \Omega_S(t)\text{Im}(a_2^*a_3)] \tag{5.141}$$

must vanish and again, we have assumed resonance $\Delta_S = \Delta_P = 0$. This requirement leads to the conditions

---

[9] Exercise 15.11 in [17] sheds more light on what "adiabatically" means in this context.

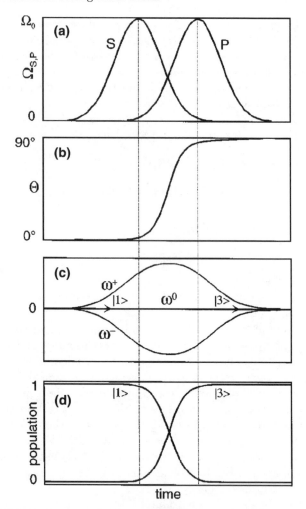

Fig. 5.35. STIRAP dynamics:(a) Pump- and Stokes-pulse, (b) angle Θ, (c) dressed eigenvalues, (d) occupation probabilities, from [51]

$$\Omega_P = -\Omega_0(t)\text{Im}[a_3^*(t)a_2(t)] \tag{5.142}$$

$$\Omega_S = \Omega_0(t)\text{Im}[a_1^*(t)a_2(t)]. \tag{5.143}$$

The two terms on the RHS of (5.141) then cancel each other. The counter-intuitive ordering of the pulses follows from the fact that a_1 is initially large and therefore also Ω_S is large compared with Ω_P. When the system is in state $|3\rangle$ the pump pulse takes over.

5.4.3 Optimal Control Theory

Optimal control theory deals with the search for external fields that steer a system into a desired state. This can e.g., be a certain vibrational excitation, which was also the goal of STIRAP, just discussed. One of the most demanding goals that can be reached with lasers is the control of a chemical reaction, however. One can e.g., try to design laser pulses in such a way that chemically bound species dissociate in a predetermined way. In a triatomic system, several different reaction channels exist. Two of them are

$$ABC \rightarrow A + BC \qquad \text{channel 1}$$

and

$$ABC \rightarrow AB + C \qquad \text{channel 2,}$$

and a laser that discriminates channel 1 in favor of channel 2 might e. g. be looked for.

In the following, we will discuss two important scenarios in the field of chemical reactions, starting with the "precursor" of the optimal control schemes, the so-called "pump-dump"-scheme and then reviewing in detail the Krotov method, which gives a mathematical prescription to find the optimal field. Finally, we will come back to the question of steering a system into a desired quantum state.

Pump-Dump Control

The so-called pump-dump method is a very intuitive way to approach the field of optimal control [52]. One tries to steer the breaking of a specific bond by first lifting the system (e.g., a collinear ABC system) onto an electronically excited state and then using the motion of the nuclei in that state in such way that the system is deexcited exactly at a time when the motion in the electronic ground state leads to dissociation in the desired channel.

To understand the physics behind the pump-dump method, we first need to look at a typical potential landscape of an ABC system, e.g., shown in Fig. 5.36. The lower surface has a local minimum and two channels, which are separated from the minimum via saddle points. The upper surface is almost harmonic. The basic idea of how to steer the reaction into a desired channel becomes clear, if we consider the classical Lissajous motion on the electronically excited surface after excitation with the pump pulse. This motion is depicted in the leftmost panel in Fig. 5.37. Now the dump-pulse arrives with a specific time delay. The Husimi transform of a typical pump-dump pulse sequence has been displayed in Fig. 1.10 of Chap. 1 already. Choosing the time delay accordingly, the Lissajous motion can be intercepted at any desired point. If it is intercepted at t_1, such that the motion on the electronic

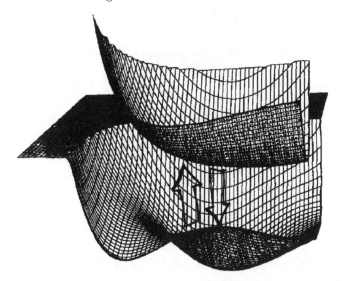

Fig. 5.36. Electronic potential curves of a collinear ABC system with pictorial representation of the pump and the dump pulse, from [53]

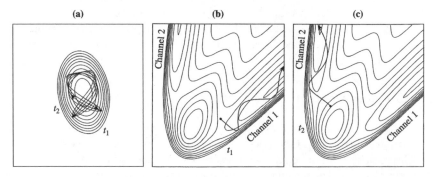

Fig. 5.37. Classical mechanical understanding of the pump-dump scenario in the ABC system: (**a**) Lissajous motion on upper surface, (**b**) classical trajectory exiting in channel 1, (**c**) classical trajectory exiting in channel 2; from [54]

ground state continues in channel 1, then the dissociation has been steered to proceed in this channel, as depicted e. g. in the middle panel of Fig. 5.37.

Quantum mechanically, the dynamics of wavefunctions and not of a single classical trajectory has to be considered. It turns out, however, that due to the harmonic nature of the excited electronic state, the physical picture of the pump-dump method stays intact [54]. The wavepacket evolves almost dispersionless on the upper surface and the description in terms of classical trajectories is sufficient. After action of the dump pulse with a time delay of e. g. 810 a.u. the wavepacket exits in channel 2 on the electronic ground state as depicted in Fig. 5.38. This case corresponds to the rightmost panel of Fig. 5.37.

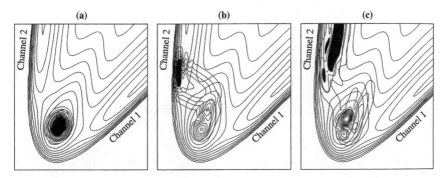

Fig. 5.38. Quantum mechanical pump-dump scenario in the HHD system for a time delay of 810 a.u.: (**a**) wave function in the electronic ground state at the initial time $t = 0$ a.u., (**b**) wave function in the electronic ground state at the time $t = 1,000$ a.u., (**c**) wave function in the electronic ground-state at the time $t = 1,200$ a.u., from [54]

Krotov Method

The pump-dump method just discussed is the precursor of modern control methods that try to achieve higher yields, i.e., to achieve the desired goal to a higher degree.

The goal can by formulated mathematically by using a projection operator $\hat{P}_\alpha$, projecting the wavefunction on the desired channel and trying to maximize

$$J_P = \lim_{T_t \to \infty} \langle \chi(T_t) | \hat{P}_\alpha | \chi(T_t) \rangle. \tag{5.144}$$

Here T_t is the total time allowed for the control process.

In order that the energy content of the field does not grow indefinitely, the functional above is usually augmented by a term

$$J_{\mathcal{E}} = \lambda \int_0^{T_t} dt |\mathcal{E}(t)|^2, \tag{5.145}$$

proportional to a Lagrange multiplier λ. Furthermore, the time-dependent Schrödinger equation is introduced again via a Lagrange multiplier $\langle \xi(t) |$ into the functional by the real term

$$J_H = 2\mathrm{Re} \int_0^{T_t} dt \langle \xi(t) | - \partial_t + \frac{\hat{H}}{i} | \chi(t) \rangle, \tag{5.146}$$

which deconstrains $\mathcal{E}$ and χ [17]. The functional to be extremized is finally given by

$$\tilde{J} \equiv J_P + J_H - J_{\mathcal{E}}. \tag{5.147}$$

To be specific, the case of a two component wavefunction $\boldsymbol{\chi} = (\chi_{\mathrm{g}}, \chi_{\mathrm{e}})$ and a corresponding 2×2 Hamilton matrix

$$\hat{\mathbf{H}} = \begin{pmatrix} \hat{H}_{\mathrm{g}} & \mu\mathcal{E}^*(t) \\ \mu\mathcal{E}(t) & \hat{H}_{\mathrm{e}} \end{pmatrix} \tag{5.148}$$

is considered in the following. After partial integration of the J_H term,

$$\tilde{J} = J_P - 2\mathrm{Re}\langle\boldsymbol{\xi}|\boldsymbol{\chi}\rangle|_0^{T_{\mathrm{t}}} + 2\mathrm{Re}\int_0^{T_{\mathrm{t}}} \mathrm{dt}\left\{\langle\boldsymbol{\xi}(t)|\frac{\hat{\mathbf{H}}}{\mathrm{i}}|\boldsymbol{\chi}(t)\rangle + \langle\dot{\boldsymbol{\xi}}|\boldsymbol{\chi}\rangle\right\} - J_\epsilon \tag{5.149}$$

is found. The variation of this expression can now be performed according to the rules that are gathered in Appendix 2.B. Extremalizing with respect to $\boldsymbol{\chi}$, i.e. the condition

$$\frac{\delta\tilde{J}}{\delta|\boldsymbol{\chi}(t)\rangle} = 0 \tag{5.150}$$

leads to the equation

$$-\mathrm{i}\langle\dot{\boldsymbol{\xi}}| = \langle\boldsymbol{\xi}|\hat{\mathbf{H}}, \tag{5.151}$$

which is a backward Schrödinger equation for the Lagrange parameter. Its final condition is found by doing the variation

$$\frac{\delta\tilde{J}}{\delta|\boldsymbol{\chi}(T_{\mathrm{t}})\rangle} = 0, \tag{5.152}$$

leading to

$$\langle\boldsymbol{\xi}(T_{\mathrm{t}})| = \langle\boldsymbol{\chi}(T_{\mathrm{t}})|\hat{P}_\alpha. \tag{5.153}$$

In addition to this equation, also the initial value equation

$$\mathrm{i}|\dot{\boldsymbol{\chi}}\rangle = \hat{\mathbf{H}}|\boldsymbol{\chi}\rangle \tag{5.154}$$

$$|\boldsymbol{\chi}(0)\rangle = |\boldsymbol{\chi}_0\rangle, \tag{5.155}$$

has to hold.

Extremalizing with respect to $\mathcal{E}^*$, i. e. the condition

$$\frac{\delta\tilde{J}}{\delta\mathcal{E}^*(t)} = 0 \tag{5.156}$$

leads to

$$\mathcal{E}(t) = \frac{-\mathrm{i}}{\lambda}[\langle\xi_{\mathrm{g}}|\mu|\chi_{\mathrm{e}}\rangle - \langle\chi_{\mathrm{g}}|\mu|\xi_{\mathrm{e}}\rangle] \tag{5.157}$$

for the field [53].

Exercise 5.5 *Show that setting the variation of $J_H - J_{\mathcal{E}}$ with respect to $\mathcal{E}^*$ equal to zero leads to the expression (5.157) for the field.*

The five equations (5.151,5.153,5.154,5.155,5.157) contain a double-sided boundary value problem. The easiest solution procedure is given by the following steps:

1. Propagate $\chi(t)$ from $t = 0$ to $t = T_t$ forward in time
2. Apply $\hat{P}_\alpha$ to $\chi(T_t)$ yielding $\xi(T_t)$
3. Propagate ξ from $t = T_t$ to $t = 0$ backward in time

The field has to be guessed, however, and does not necessarily fulfill the equation coming out of the variation procedure! Therefore, the scheme above has to be augmented by an iterative procedure, due to Krotov [55]:

1. Choose an initial field $\mathcal{E}_0(t)$
2. Propagate $\chi(t)$ under $\mathcal{E}_0(t)$ forward in time
3. Projection of $\chi(T_t)$ gives $\xi(T_t)$
4. Propagate ξ backward in time
5. Commonly propagate $\xi(t)$ (with the old field) and $\chi(t)$ with the new instantaneously calculated field

$$\mathcal{E}_1(t) = \frac{-i}{\lambda}[\langle\xi_g^0|\mu|\chi_e^1\rangle - \langle\chi_g^1|\mu|\xi_e^0\rangle] \tag{5.158}$$

   forward in time
6. Project $\chi(T_t)$ and continue the procedure until convergence is achieved.

The propagation of $\xi(t)$ forward in time seems to be superfluous, because the result is already known. Keeping the wavefunction in computer memory would be barely possible for most cases of interest, however, and therefore it is cheaper to calculate $\xi(t)$ once more. In general, the propagated wavefunction has amplitude in both the desired and the undesired channel, see e.g., figure 2f in [53]. The above procedure iterates the field in such way that the undesired portion of the wavefunction is minimized. If this minimum is an absolute or a local one is a question that goes far beyond the scope of this book. The method just laid out goes back to Krotov. Other methods to solve for the optimal field have been devised, however, see e. g. [17] and [56] and the references therein.

As an example, let us review results that were obtained for an ABC system. As the initial guess for the electric field in the iterative process, a pump-dump pulse as depicted in Fig. 1.10 has been used. One can try to steer the reaction either into channel 1 or into channel 2. In the first case, the resulting optimal field is still rather similar to the original pump-dump pulse [17], whereas in the second case the field displayed in Fig. 5.39 is resulting. Around 50 iterations are typically necessary to converge the results.

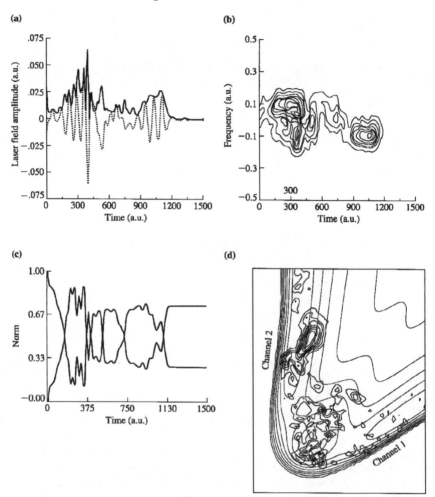

Fig. 5.39. Steering the breakup of the ABC system into the second channel: (a) Optimal field, (b) Husimi transform of the field, (c) norm of the wavefunction in the ground and excited state, (d) final wavefunction, from [17]

Optimally Controlled Excitation of Quantum States

Optimal control schemes do not only work for the breakup reaction just considered. They have been shown to be applicable also for the case of vibrational excitation.

In [25] optimal control theory has e. g. been applied with the objective to steer a Morse oscillator, representing a CH-stretch , with Morse parameters $D_e = 0.199$, $R_e = 1.5$, and $\alpha = 0.9386$ into a specific excited state $|n_T\rangle$. The projection operator therefore is given by

$$\hat{P}_T = |n_T\rangle\langle n_T|. \tag{5.159}$$

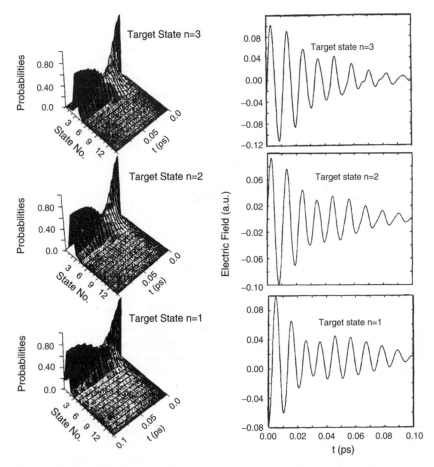

Fig. 5.40. Steering a Morse oscillator into selected excited states via optimal control theory, left panels show probabilities as a function of time and quantum number, right panels show the corresponding fields; from [25]

The dipole moment was assumed to be of Mecke form with the parameters $\mu_0 = 1.76$ a.u. and $R^* = 1$ a.u.. The result of the optimization starting from the vibrational ground state and fixing the final time to be $T_t = 0.1$ ps are shown in Fig. 5.40.

Optimal control theory has been applied in a lot of other physical systems. One out of many other examples is the control of *cis-trans* isomerization [57].

5.4.4 Genetic Algorithms

The theory of optimal control of the last section rests on the availability of analytically (or numerically) given potential energy surfaces and on the validity of the underlying Born-Oppenheimer approximation. Both requirements

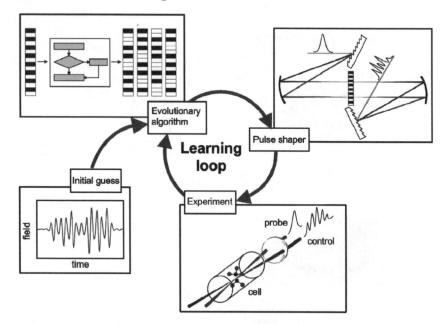

Fig. 5.41. Principal setup of an "analog computer" for feedback control. The control pulse excites the system and the probe pulse measures the outcome (the measurement could also be performed e.g., by a mass spectrometer), which is fed into the evolutionary algorithm; a "good" initial guess helps to achieve convergence quickly, from [58]

may be violated, and even the Hamiltonian might not be known, however, and alternative control schemes are sought for.

A recent development in the field of control, therefore, is the application of genetic algorithms. Their application is based on an experimental "analog computer". The system to be controlled is exposed to a laser whose temporal shape can be varied. By a feedback mechanism, a digital computer using a genetic (evolutionary) algorithm can vary the field iteratively in such a way that the desired goal is reached. The principal setup is displayed in Fig. 5.41.

The first theoretical study that showed the feasibility of such an approach is due to Judson and Rabitz [59]. These authors have shown that it is based on the following three paradigms

- "Survival of the fittest"
- Crossover
- Mutation

As an example the transition from the $\nu = j = 0$ ground vibrational state to the rotationally excited vibrational ground state $\nu = 0, j' = 3$ of the KCl molecule was investigated. An "individuum" of the genetic algorithm is a specific pulse sequence. Its initial gene consists of $N_{\text{gene}} = 128$ entries of random numbers, uniformly distributed between zero and one and scaled to a maximum of $5\,\mathrm{kV\,cm^{-1}}$. The total number of individuals was $N_{\mathrm{pop}} = 50$. The first paradigm can be tested by introducing the "cost-function"

$$\sum_j (\delta_{jj'} - \rho_j)^2,$$

where ρ_j is the occupation probability of state j. Indiviuals can be ranked according to their fitness. The highest ranked ones were taken over to the next generation without change, whereas the other ones had to undergo crossover and a small probability of mutation. In the upper panel of Fig. 5.42, the decay of the cost function as a function of generation is displayed for the average population, as well as for the best individuum. Furthermore, in the lower panel of that Figure, the spectrum corresponding to the best gene is shown. It displays maxima at the resonant transitions between the rotational states $j = 0, 1, 2, 3$. It is important to stress that the spectral information was not input into the generation of the optimal field but was found by the learning loop.

An experimental realization of control based on evolutionary algorithms was performed by the Gerber group [60]. The goal of the experiment was to steer the photo-fragmentation of $CpFe(CO)_2Cl$ into a desired channel. A pulse shaper that allows to split the laser light into 128 spectral components and vary them separately has been used. Contrast ratios of about 5 have been achieved.

5.4.5 Toward Quantum Computing with Molecules

A recent new development in the field of laser molecule interaction is the realization of quantum logic operations with the help of molecular vibrational states. We will not deal with that exciting new field in much detail but will discuss the realization of the basic ingredient of every setup used for computing: the flipping of a bit. As we will see, anharmonic vibrational modes have to be used to this end. Using the OH diatomic, this has e.g., been shown by Babikov [61] and by Cheng and Brown [62].

The flipping of a bit is based on the realization of the NOT-Operation. In a two-level system this corresponds to the complete transfer of population from level 0 to level 1 or vice versa

$$\mathrm{NOT}|0\rangle = |1\rangle \qquad\qquad (5.160)$$
$$\mathrm{NOT}|1\rangle = |0\rangle \qquad\qquad (5.161)$$

Each deviation from the complete population transfer reduces the so-called fidelity, defined as the occupation probability of the initially unpopulated level. Unfortunately, a harmonic oscillator cannot be controlled to switch completely to a desired state, because of the equidistance of its levels. One has to choose two levels of an anharmonic system, as e.g., the Morse potential of section 5.1.2 in order to realize the NOT operation.

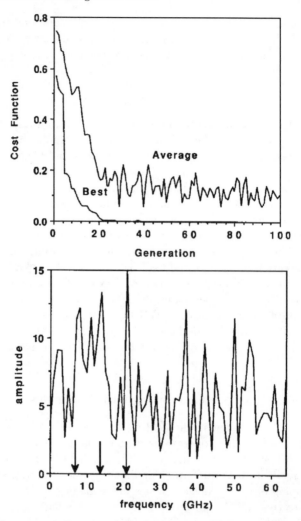

Fig. 5.42. *Upper panel*: Cost function of the average population and the best gene; *lower panel*: Spectrum of the optimal pulse (*arrows* indicate positions of resonant transitions between rotational sublevels of the vibrational ground state of the KCl molecule), from [59]

Exercise 5.6 *Derive the maximal probability to populate the n-th excited state of a harmonic oscillator by using an external field in length gauge and starting from the ground state χ_0.*

a) *Calculate the time-dependent wavefunction $\chi_0(x, t)$ under the influence of the external field.*

b) *Determine the overlap*

$$a_{n0}(t) = \langle \chi_n | \chi_0(t) \rangle$$

by using $\int dx \exp[-(x-y)^2]H_n(x) = \pi^{1/2}(2y)^n$ with the Hermite polynomial $H_n(x)$.

c) Show that the maximum of the absolute value $|a_{n0}(t)|^2$ is given by $n^n e^{-n}/n!$

As a specific example, the driven OH stretch with the Morse parameters $D_e = 0.1994$, $R_e = 1.821$, and $\alpha = 1.189$ in atomic units and the Mecke parameters $\mu_0 = 1.634$ a.u. and $R^* = 1.134$ a.u. (see also caption of Fig. 5.13) have been used. Although the anharmonicity constant of that molecule is fixed in nature, it can be viewed as a parameter in theoretical considerations, see e.g., [62]. These authors have looked at the fidelity

$$P_{10}(T_t) = |\langle 1|0(T_t)\rangle|^2$$

with $T_t = 750$ fs as a function of anharmonicity, and found the results reproduced in Fig. 5.43. Two different results are shown there. First, the system has been exposed to an optimal control pulse, in close analogy to the work of Shi and Rabitz [25], and second, to a simple π-pulse, we know already from Chap. 3. For large anharmonicity, it can be seen that the π-pulse is superior to the "optimal" pulse! Furthermore, the statement that the harmonic oscillator cannot be controlled to 100% can be read off from the results at small anharmonicity.

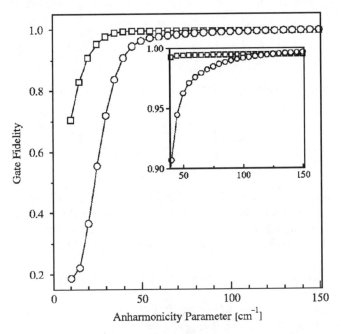

Fig. 5.43. Fidelity of the NOT gate as a function of the anharmonicity $\omega_e x_e$ for an optimal pulse (*squares*) and for the π-pulse (*circles*), adapted from [62]

5.A Relative and Center of Mass Coordinates for H_2^+

To derive the Hamiltonian for H_2^+ in a laser field, we follow Hiskes's general treatment of diatomic molecules [63]. Specializing to the single electron case, the Hamiltonian in length gauge is first expressed by using the coordinates of the nuclei R_a, R_b and of the electron, r_e, according to

$$\hat{H}_{\text{tot}} = -\frac{1}{2}\{\Delta_a/M_p + \Delta_b/M_p + \Delta_e\} + V_1 - \mathcal{E}(t)[z_a + z_b - z_e], \quad (5.162)$$

with the proton mass M_p, and where V_1 contains all Coulomb interaction terms, the laser is polarized in z direction, and we have used atomic units.

Now center of mass and relative coordinates

$$R_S = \frac{M_p R_a + M_p R_b + r_e}{M_S}$$

$$R = R_a - R_b$$

$$r_i = -\frac{R_a + R_b}{2} + r_e$$

are introduced. $M_S = 2M_p + 1$ is the total mass of the system (in a.u.) and the coordinate of the electron is measured relative to the center of mass of the nuclei.

In matrix form, the old and the new coordinates are related by

$$\begin{pmatrix} R_S \\ R \\ r_i \end{pmatrix} = \begin{pmatrix} \frac{M_p}{M_S} & \frac{M_p}{M_S} & \frac{1}{M_S} \\ 1 & -1 & 0 \\ -1/2 & -1/2 & 1 \end{pmatrix} \begin{pmatrix} R_a \\ R_b \\ r_e \end{pmatrix}. \quad (5.163)$$

With the help of the inverse matrix the back transformation can be derived which amounts to

$$\begin{pmatrix} R_a \\ R_b \\ r_e \end{pmatrix} = \begin{pmatrix} 1 & 1/2 & -\frac{1}{M_S} \\ 1 & -1/2 & -\frac{1}{M_S} \\ +1 & 0 & \frac{2M_p}{M_S} \end{pmatrix} \begin{pmatrix} R_S \\ R \\ r_i \end{pmatrix}. \quad (5.164)$$

Finally not only the old coordinates but also their time derivatives in the classical form of the Hamiltonian are expressed in terms of the new coordinates. It turns out that the center of mass with charge e moves "freely" in the electrical field [63]. The relative motion, however, is governed by the Hamiltonian

$$\hat{H}_{\text{rel}} = -\frac{1}{2}\{\Delta_R/M_r + \Delta_i/m_i\} + V_1 + \mathcal{E}(t)\left[1 + 1/M_S\right]z_i, \quad (5.165)$$

where the reduced masses $M_r = M_p/2$ and $m_i = \frac{2M_p}{M_S}$ have been introduced and the Coulomb interaction is expressed in terms of the relative coordinates.

In the case of H_2^+, no kinetic couplings between the different degrees of freedom are present nor does the field couple to the relative coordinate of the nuclei, which is certainly not true in the general case [63].

5.B Perturbation Theory for Two Coupled Surfaces

In the case of a laser driven two level system with a 2×2 (matrix-) Hamiltonian

$$\hat{\mathbf{H}} = \hat{\mathbf{H}}_0 + \hat{\mathbf{W}}(t) \tag{5.166}$$

perturbation theory is best performed in the interaction picture of Sect. 2.2.3. In first order and after a back transformation to the Schrödinger picture

$$|\boldsymbol{\chi}(t)\rangle = e^{-i\hat{\mathbf{H}}_0 t}|\boldsymbol{\chi}(0)\rangle + \frac{1}{i}\int_0^t dt' e^{-i\hat{\mathbf{H}}_0(t-t')}\hat{\mathbf{W}}(t')e^{-i\hat{\mathbf{H}}_0 t'}|\boldsymbol{\chi}(0)\rangle \tag{5.167}$$

can be written for the vector valued wavefunction in atomic units.

Invoking the Born-Oppenheimer approximation, we assume that the unperturbed Hamiltonian $\hat{\mathbf{H}}_0$ has only diagonal elements. Furthermore, the initial wavefunction shall be restricted to the electronic ground state

$$|\boldsymbol{\chi}(0)\rangle = \begin{pmatrix} |\chi_g(0)\rangle \\ 0 \end{pmatrix}. \tag{5.168}$$

Under the influence of the perturbation (having only off diagonal elements), the component of the wavefunction

$$|\boldsymbol{\chi}(t)\rangle = \begin{pmatrix} |\chi_g(t)\rangle \\ |\chi_e(t)\rangle \end{pmatrix} \tag{5.169}$$

in the excited electronic state as a function of time is the desired quantity. Under the assumptions mentioned above, for this quantity the golden rule expression in position representation

$$\chi_e(\boldsymbol{R}, t) = \frac{1}{i}\int_0^t dt' e^{-i\hat{H}_e(t-t')}\hat{W}_{eg}(\boldsymbol{R}, t')e^{-i\hat{H}_g t'}\chi_g(\boldsymbol{R}, 0) \tag{5.170}$$

is found. Here the definitions

$$\hat{W}_{eg}(\boldsymbol{R}, t) = \langle e(\boldsymbol{R})|\hat{W}(t)|g(\boldsymbol{R})\rangle \qquad \hat{H}_j = \langle j(\boldsymbol{R})|\hat{H}_0|j(\boldsymbol{R})\rangle \tag{5.171}$$

of the matrix elements of the Hamiltonian[10] have been introduced, and j can either be e(xited) or g(round). The physical interpretation of this final result is straightforward. The system propagates for a time t' on the lower surface is then lifted to the upper surface and propagates there until the final time t. All the possibilities to split the time interval $[0, t]$ have to be integrated over.

---

[10] Please note that the matrix elements are still operators (as indicated by the hat), due to the fact that the integrations in (5.171) are only over electronic coordinates.

5.C Reflection Principle of Photodissociation

The dynamical reflection principle plays a major role for the interpretation of the photoelectron spectrum in a pump-probe experiment. It has an analog in the field of photodissociation. For example, in [18] it is shown that the absorption spectrum in photodissociation is given by the Fourier transform of the auto-correlation function of a ground state wavepacket, approximated by a Gaussian with width parameter α_R, that is instantaneously lifted to the excited state, where it is evolving in time. In order to be able to do analytic calculations, this antibinding surface is approximated by a straight line

$$V(R) \approx V_e - V_R(R - R_e). \tag{5.172}$$

Using the short-time approximation (i.e., neglecting the kinetic energy) the wavefunction on the antibinding surface is given by

$$\chi(R,t) \sim e^{-i[V_e - V_R(R-R_e)]t} e^{-\alpha_R(R-R_e)^2} \tag{5.173}$$

The Fourier transformation of the auto-correlation $c(t) = \langle \chi(0)|\chi(t)\rangle$ can be done analytically, yielding

$$\sigma(E) \sim \frac{e^{-2\beta(E-V_e)^2}}{V_R}, \tag{5.174}$$

with $\beta = (V_R^2/\alpha_R)^{-1}$. The same result can also be obtained by a purely classical calculation [18].

The maximum of the spectrum is at $E = V_e$ and its width is proportional to the negative slope of the antibinding surface. Both facts can be seen by looking at Fig. 5.44. There it is shown that the reflection of the square of the initial wavepacket at the antibinding surface yields the spectrum.

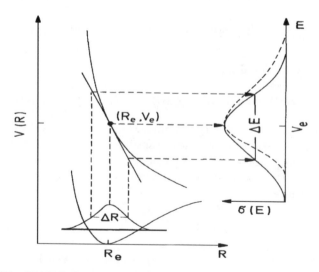

Fig. 5.44. Reflection principle of photodissociation; from [18]

5.D The Undriven Double Well Problem

Figure 5.45 shows the unperturbed double well potential

$$V_{DW}(x) \equiv -\frac{1}{4}x^2 + \frac{1}{64D}x^4 \qquad (5.175)$$

in the units introduced in Sect. 5.4.1 for $D = 2$, including the five energy eigenvalues which lay below the barrier.

The coherent tunneling of a particle in the double well potential emerges by considering an initial state that is a superposition of the two lowest eigenfunctions, $\chi_1(x), \chi_2(x)$, depicted in Fig. 5.46, with the energies E_1, E_2.[11] At time $t = 0$ this leads to a state that is localized in the left well

$$\chi_l(x,0) = \frac{1}{\sqrt{2}}\left[\chi_1(x) - \chi_2(x)\right]. \qquad (5.176)$$

In the case $D \rightarrow \infty$ it is identical to the ground state of the harmonic approximation to the left well. Its absolute value has the time evolution

$$|\chi_l(x,t)|^2 = \frac{1}{2}\left\{|\chi_1(x)|^2 + |\chi_2(x)|^2 - 2\chi_1(x)\chi_2(x)\cos[(E_2 - E_1)t]\right\}. \quad (5.177)$$

Defining the tunneling splitting as

$$\Delta = E_2 - E_1, \qquad (5.178)$$

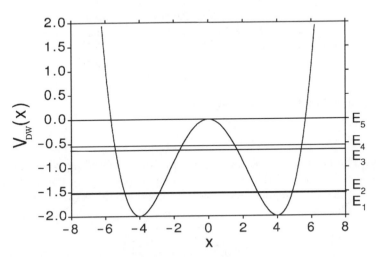

Fig. 5.45. Symmetric double well potential with five energy eigenvalues for $D = 2$

---

[11] Please note that in this appendix $\chi_j(x)$ denotes the jth eigenfunction in the same electronic state (the double well).

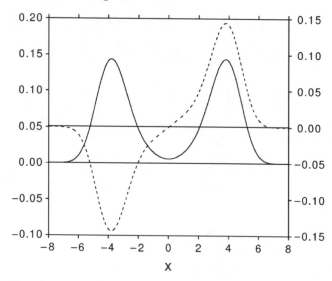

Fig. 5.46. Eigenfunctions of the lowest two Eigenvalues for $D = 2$. *Solid line:* $\chi_1(x)$, *dashed line:* $\chi_2(x)$

Table 5.3. Numerical and semiclassical (index s) results for the tunneling splitting as a function of the barrier height; from [42]

D	Δ	Δ_{s}	$\frac{\Delta_{\mathrm{s}}-\Delta}{\Delta}$
1	2.392×10^{-2}	3.082×10^{-2}	28.8%
1.5	2.262×10^{-3}	2.623×10^{-3}	15.9%
2	1.895×10^{-4}	2.104×10^{-4}	11.0%
2.5	1.507×10^{-5}	1.635×10^{-5}	8.5%
3	1.164×10^{-6}	1.244×10^{-6}	6.8%

the corresponding tunneling time

$$T_{\mathrm{tu}} = \frac{2\pi}{\Delta} \tag{5.179}$$

follows. The eigenvalues of the time-independent Schrödinger equation with a quartic potential and thus also the tunneling splitting are not available exactly analytically. Using a semiclassical approximation, Δ can be determined, however. The result of such a calculation is [64]

$$\Delta_{\mathrm{s}} = 8\sqrt{\frac{2D}{\pi}} \exp\left(-\frac{16D}{3}\right), \tag{5.180}$$

depending exponentially on the dimensionsless barrier height. In Table 5.3 some values of Δ for different barrier heights can be found.

5.E The Quantum Mechanical Adiabatic Theorem

To derive the adiabatic theorem in quantum theory, let us consider a system with discrete levels, whose state vector is given by

$$|\boldsymbol{\Psi}(t)\rangle = \begin{pmatrix} |\psi_1(t)\rangle \\ |\psi_2(t)\rangle \\ \vdots \end{pmatrix}. \tag{5.181}$$

In case of a time-dependent perturbation, the Hamilton matrix is given by

$$\mathbf{H}(t) = \begin{pmatrix} E_1 & V_{12}(t) & \cdots \\ V_{21}(t) & E_2 & \cdots \\ \vdots & \vdots & \vdots \end{pmatrix}. \tag{5.182}$$

Clearly, an eigenstate stays an eigenstate without the external perturbation. Even in the presence of a slowly changing perturbation an analogous statement holds, however.

To show this, one first defines a unitary transformation, diagonalizing the instantaneous Hamiltonian

$$\mathbf{U}^{-1}(t)\mathbf{H}(t)\mathbf{U}(t) = \mathbf{D}(t) \tag{5.183}$$

The transformed state vector is given by

$$|\boldsymbol{\Psi}'(t)\rangle = \mathbf{U}^{-1}|\boldsymbol{\Psi}(t)\rangle. \tag{5.184}$$

It fulfills the time-dependent Schrödinger equation

$$i\hbar\dot{\boldsymbol{\Psi}}(t)\rangle = \mathbf{D}|\boldsymbol{\Psi}'(t)\rangle - i\hbar\mathbf{U}^{-1}\dot{\mathbf{U}}|\boldsymbol{\Psi}'(t)\rangle \tag{5.185}$$

If the Hamilton matrix $\mathbf{H}$ is slowly time-dependent, then also $\mathbf{U}$ depends only weakly on time and the second term on the right hand side of the equation above can be neglected. An eigenfunction of the original Hamiltonian thus stays an eigenfunction instantaneous Hamiltonian. In [65] it has been shown that for periodically driven systems the adiabatic theorem has to be modified.

Finally, it is worthwhile to mention that by choosing the perturbation in such a way that the Hamiltonian switches from a simple to a complex one, the eigenstates of the complex Hamiltonian can be gained numerically [66]. In addition, an application to two-level systems has been given in [67].

Notes and Further Reading

The authoritative reference on the hydrogen molecular ion is the classic book by Slater [1]. More general material on molecular spectra and molecular structure can be found in the textbook by Bransden and Joachain [2]. Spectroscopic constants that can be used e. g. for the generation of analytic Morse potentials for diatomic molecules are contained in [7]. Further information on diatomic molecules is available in the references given in [68]. A modern exposition of quantum chemistry which is the main theoretical tool for the calculation of electronic potential surfaces is given in the book by Szabo and Ostlund [3].

Reviews covering both theoretical as well as experimental facts on the dynamics of H_2^+ in intense laser fields are given in [8,69]. More information on molecules in laser fields can also be found in the book edited by Bandrauk [70]. The review by Posthumus [69] contains an insightful discussion of field dressed states of H_2^+ and their use to explain phenomena like molecular stabilization (bond hardening) and bond softening.

The transformation from adiabatic to diabatic states and the nonuniqueness of that transformation are e.g., discussed in chapter 15.2 of [18]. Additional material on this transformation can be found in chapter 12.2 of [17]. Nonadiabatic molecular dynamics can be tackled in many different ways. Further information on that topic with additional references can be found for the case without an external laser in [71] and with an external laser in [72]. The semiclassical initial value method applied to the problem of coupled surfaces is reviewed in [73]. Reviews of theoretical and experimental approaches to femtosecond chemistry are collected in [74].

Driven quantum tunneling is reviewed in depth in [75] and STIRAP is discussed in greater detail than here e. g. in the books by Rice and Zhao [56] and by Tannor [17]. Some experimental aspects on STIRAP can be found in the overview article by Bergmann et al. [51]. The formulation of most of the section on optimal control is based on chapter 16 of [17]. Tannor's book as well as [56] and [76] contain a wealth of additional material on the coherent control of quantum dynamics.

References

1. J.C. Slater, *Quantum Mechanics of Molecules and Solids*, vol. 1 (McGraw-Hill, New York, 1963)
2. B.H. Bransden, C.J. Joachain, *Physics of Atoms and Molecules*, 2nd edn. (Pearson Education, Harlow, 2003)
3. A. Szabo, N.S. Ostlund, *Modern Quantum Chemistry* (Dover, Mineola, 1996)
4. B.N. Finkelstein, G.E. Horowitz, Z. Phys. **48**, 118 (1928)
5. P.M. Morse, Phys. Rev. **34**, 57 (1929)
6. D. ter Haar, Phys. Rev. **70**, 222 (1946)
7. G. Herzberg, *Molecular Spectra and Molecular Structure: I. Spectra of Diatomic Molecules*, 2nd edn. (Krieger, Malabar, 1989)
8. A. Giusti-Suzor, F.H. Mies, L.F. DiMauro, E. Charron, B. Yang, J. Phys. B **28**, 309 (1995)
9. S. Chelkowski, T. Zuo, A.D. Bandrauk, Phys. Rev. A **46**, R5342 (1992)
10. S. Chelkowski, T. Zuo, O. Atabek, A.D. Bandrauk, Phys. Rev. A **52**, 2977 (1995)
11. M. Uhlmann, T. Kunert, R. Schmidt, Phys. Rev. A **72**, 045402 (2005)

12. J.D. Jackson, *Klassische Elektrodynamik* (Walter de Gruyter, Berlin, 1983)
13. T. Zuo, A.D. Bandrauk, Phys. Rev. A **52**, R2511 (1995)
14. A.D. Bandrauk, in *Molecules in Laser Fields*, ed. by A.D. Bandrauk (Dekker, New York, 1994), Chap. 1, pp. 1–69
15. B. Feuerstein, U. Thumm, Phys. Rev. A **67**, 043405 (2003)
16. A. Palacios, H. Bachau, F. Martin, Phys. Rev. Lett. **96**, 143001 (2006)
17. D.J. Tannor, *Introduction to Quantum Mechanics: A Time-dependent Perspective* (University Science Books, Sausalito, 2007)
18. R. Schinke, *Photodissociation Dynamics* (Cambridge University Press, Cambridge, 1993)
19. R.B. Walker, R.K. Preston, J. Chem. Phys. **67**, 2017 (1977)
20. R. Mecke, Z. Elektrochemie **54**, 38 (1950)
21. M.J. Davis, R.E. Wyatt, Chem. Phys. Lett. **86**, 235 (1982)
22. W. Jakubetz, J. Manz, V. Mohan, J. Chem. Phys. **90**, 3683 (1989)
23. S. Chelkowski, A.D. Bandrauk, P.B. Corkum, Phys. Rev. Lett. **65**, 2355 (1990)
24. S. Krempl, T. Eisenhammer, A. Hübler, G. Mayer-Kress, P.W. Milonni, Phys. Rev. Lett. **69**, 430 (1992)
25. S. Shi, H. Rabitz, J. Chem. Phys. **92**, 364 (1990)
26. B.M. Garraway, K.A. Suominen, Rep. Prog. Phys. **58**, 365 (1995)
27. J.C. Tully, J. Chem. Phys. **93**, 1061 (1990)
28. H.D. Meyer, W.H. Miller, J. Chem. Phys. **72**, 2272 (1980)
29. G. Stock, M. Thoss, Phys. Rev. Lett. **78**, 578 (1997)
30. J.J. Sakurai, *Modern Quantum Mechanics* (Addison-Wesley, Reading, 1994)
31. K.A. Suominen, B.M. Garraway, S. Stenholm, Phys. Rev. A **45**, 3060 (1992)
32. I.S. Gradshteyn, I.M. Ryzhik, *Table of Integrals Series and Products*, 5th edn. (Academic, New York, 1994)
33. F. Grossmann, Phys. Rev. A **60**, 1791 (1999)
34. A. Assion, M. Geisler, J. Helbing, V. Seyfried, T. Baumert, Phys. Rev. A **54**, R4605 (1996)
35. C. Meier, Ph.D. Thesis, *Theoretische Untersuchungen zur Photoelektronenspektroskopie kleiner Moleküle mit kurzen und intensiven Laserpulsen* (Universität Freiburg, 1995)
36. V. Engel, Comput. Phys. Commun. **63**, 228 (1991)
37. V. Engel, H. Metiu, J. Chem. Phys. **100**, 5448 (1994)
38. S.O. Williams, D.G. Imre, J. Phys. Chem. **92**, 6648 (1988)
39. M. Dantus, M.J. Rosker, A.H. Zewail, J. Chem. Phys. **87**, 2395 (1987)
40. F. Hund, Z. Phys. **43**, 805 (1927)
41. F. Grossmann, T. Dittrich, P. Jung, P. Hänggi, Phys. Rev. Lett. **67**, 516 (1991)
42. F. Grossmann, Ph.D. Thesis, *Der Tunneleffekt in periodisch getriebenen Quantensystemen* (Universität Augsburg, 1992)
43. F. Grossmann, P. Hänggi, Europhys. Lett. **18**(1), (1992)
44. J.H. Shirley, Phys. Rev. **138**, B979 (1965)
45. M. Abramowitz, I.A. Stegun, *Handbook of Mathematical Functions* (Dover, New York, 1965)
46. Y. Kayanuma, Phys. Rev. A **50**, 843 (1994)
47. G.D. Valle, M. Ornigotti, E. Cianci, V. Foglietti, P. Laporta, S. Longhi, Phys. Rev. Lett. **98**, 263601 (2007)
48. E. Kierig, U. Schnorrberger, A. Schietinger, J. Tomkovic, M.K. Oberthaler, Phys. Rev. Lett. **100**, 190405 (2008)

49. V. Roudnev, B.D. Esry, I. Ben-Itzhak, Phys. Rev. Lett. **93**, 163601 (2004)
50. M.F. Kling, C. Siedschlag, A.J. Verhoef, J.I. Khan, M. Schultze, T. Uphues, Y. Ni, M. Uiberacker, M. Drescher, F. Krausz, M.J.J. Vrakking, Science **312**, 246 (2006)
51. K. Bergmann, H. Theuer, B.W. Shore, Rev. Mod. Phys. **70**, 1003 (1998)
52. D.J. Tannor, S.A. Rice, Adv. Chem. Phys. **70**, 441 (1988)
53. R. Kosloff, S.A. Rice, P. Gaspard, S. Tersigni, D.J. Tannor, Chem. Phys. **139**, 201 (1986)
54. D.J. Tannor, R. Kosloff, S.A. Rice, J. Chem. Phys. **85**, 5805 (1986)
55. V.F. Krotov, I.N. Feldman, Eng. Cybernetics **21**, 123 (1984)
56. S.A. Rice, M. Zhao, *Optical Control of Molecular Dynamics* (Wiley, New York, 2000)
57. F. Grossmann, L. Feng, G. Schmidt, T. Kunert, R. Schmidt, Europhys. Lett. **60**, 201 (2002)
58. D. Zeidler, S. Frey, K.L. Kompa, M. Motzkus, Phys. Rev. A **64**, 023420 (2001)
59. R.S. Judson, H. Rabitz, Phys. Rev. Lett. **68**, 1500 (1992)
60. A. Assion, T. Baumert, M. Bergt, T. Brixner, B. Kiefer, V. Seyfried, M. Strehle, G. Gerber, Science **282**, 919 (1998)
61. D. Babikov, J. Chem. Phys. **121**, 7577 (2004)
62. T. Cheng, A. Brown, J. Chem. Phys. **124**, 034111 (2006)
63. J.R. Hiskes, Phys. Rev. **122**, 1207 (1961)
64. U. Weiss, W. Häffner, Phys. Rev. D **27**, 2916 (1983)
65. D.W. Hone, R. Ketzmerick, W. Kohn, Phys. Rev. A **56**, 4045 (1997)
66. D. Kohen, D.J. Tannor, J. Chem. Phys. **98**, 3168 (1993)
67. A. Emmanouilidou, X.G. Zhao, P. Ao, Q. Niu, Phys. Rev. Lett. **85**, 1626 (2000)
68. P. Passarinho, M.L. da Silva, J. Mol. Spectros. **236**, 148 (2006)
69. J.H. Posthumus, Rep. Prog. Phys. **67**, 623 (2004)
70. A.D. Bandrauk (ed.), *Molecules in Laser Fields* (Dekker, New York, 1994)
71. U. Saalmann, R. Schmidt, Z. Phys. D **38**, 153 (1996)
72. T. Kunert, R. Schmidt, Eur. Phys. J. D **25**, 15 (2003)
73. M. Thoss, H. Wang, Annu. Rev. Phys. Chem. **55**, 299 (2004)
74. J. Manz, L. Wöste (eds.) *Femtosecond Chemistry*, vols. 1 and 2 (VCH, Weinheim, 1995)
75. M. Grifoni, P. Hänggi, Phys. Rep. **304**, 229 (1998)
76. P. Brumer, M. Shapiro, *Principles of the Quantum Control of Molecular Processes* (Wiley-VCH, Berlin, 2003)

Index

Springer Series on
ATOMIC, OPTICAL, AND PLASMA PHYSICS

Springer Series on
ATOMIC, OPTICAL, AND PLASMA PHYSICS

Printing: Krips bv, Meppel, The Netherlands
Binding: Stürtz, Würzburg, Germany